Cambridge Elements

Elements in the Philosophy of Biology
edited by
Grant Ramsey
KU Leuven
Michael Ruse
Florida State University

EVOLUTION AND DEVELOPMENT

Conceptual Issues

Alan C. Love
University of Minnesota

CAMBRIDGE
UNIVERSITY PRESS

Shaftesbury Road, Cambridge CB2 8EA, United Kingdom

One Liberty Plaza, 20th Floor, New York, NY 10006, USA

477 Williamstown Road, Port Melbourne, VIC 3207, Australia

314–321, 3rd Floor, Plot 3, Splendor Forum, Jasola District Centre, New Delhi – 110025, India

103 Penang Road, #05–06/07, Visioncrest Commercial, Singapore 238467

Cambridge University Press is part of Cambridge University Press & Assessment, a department of the University of Cambridge.

We share the University's mission to contribute to society through the pursuit of education, learning and research at the highest international levels of excellence.

www.cambridge.org
Information on this title: www.cambridge.org/9781009468022

DOI: 10.1017/9781108616751

© Alan C. Love 2024

This publication is in copyright. Subject to statutory exception and to the provisions of relevant collective licensing agreements,with the exception of the Creative Commons version the link for which is provided below,no reproduction of any part may take place without the written permission of Cambridge University Press & Assessment.

An online version of this work is published at doi.org/10.1017/9781108616751 under a Creative Commons Open Access license CC-BY-NC-ND 4.0 which permits re-use, distribution and reproduction in any medium for non-commercial purposes providing appropriate credit to the original work is given. You may not distribute derivative works without permission. To view a copy of this license, visit https://creativecommons.org/licenses/by-nc-nd/4.0

When citing this work, please include a reference to the DOI 10.1017/9781108616751

First published 2024

A catalogue record for this publication is available from the British Library.

ISBN 978-1-009-46802-2 Hardback
ISBN 978-1-108-72752-5 Paperback
ISSN 2515-1126 (online)
ISSN 2515-1118 (print)

Cambridge University Press & Assessment has no responsibility for the persistence or accuracy of URLs for external or third-party internet websites referred to in this publication and does not guarantee that any content on such websites is, or will remain, accurate or appropriate.

Evolution and Development

Conceptual Issues

Elements in the Philosophy of Biology

DOI: 10.1017/9781108616751
First published online: February 2024

Alan C. Love
University of Minnesota
Author for correspondence: Alan C. Love, aclove@umn.edu

Abstract: The intersection of development and evolution has always harbored conceptual issues, but many of these are on display in contemporary evolutionary developmental biology (evo-devo). These issues include: (1) the precise constitution of evo-devo, with its focus on both the evolution of development and the developmental basis of evolution, and how it fits within evolutionary theory; (2) the nature of evo-devo model systems that comprise the material of comparative and experimental research; (3) the puzzle of how to understand the widely used notion of "conserved mechanisms"; (4) the definition of evolutionary novelties and expectations for how to explain them; and (5) the demand of interdisciplinary collaboration that derives from investigating complex phenomena at key moments in the history of life, such as the fin–limb transition. This Element treats these conceptual issues with close attention to both empirical detail and scientific practice to offer new perspectives on evolution and development. This Element is also available as Open Access on Cambridge Core.

Keywords: evo-devo, homology, interdisciplinarity, models, novelty

ISBNs: 9781009468022 (HB), 9781108727525 (PB), 9781108616751 (OC)
ISSNs: 2515-1126 (online), 2515-1118 (print)

Contents

1 What is Evo-Devo?
1.1 Some Historical Perspective

When a student learns about the American Revolution of the eighteenth century in secondary education, they often encounter moments that encapsulate the beginning, end, or climax of this complex historical event. Famous among these is "the shot heard round the world," which picks out the 1775 battle of Concord (Massachusetts) as the touchstone of open hostilities between British soldiers and colonialists that marks the formal start of the American War of Independence. There were certainly "firsts" that day, including first shots fired by colonial minutemen because of explicit orders and the first British fatalities. And the phrase is memorable, deriving from Ralph Waldo Emerson's "Concord Hymn" written sixty-two years later when the international significance of the American Revolution was more recognizable. However, any historian working on this period will tell you that the variety of events over many years leading up to this battle, the battle itself, and subsequent events yield a more tangled tale. The secondary student will no doubt be aided by Emerson's slogan in preparing for an exam, but a deeper understanding of the initiation of the American Revolution and its complicated architecture requires more (McDonnell 2016).

References to the origin of evolutionary developmental biology (evo-devo) have a memorable moment that encapsulates its contemporary beginnings for many biologists. This is the discovery of the evolutionary conservation of ~180 base pairs of nucleotide sequence (the "homeobox") responsible for the ~60 amino acid "homeodomain" region of DNA-binding proteins – transcription factors – that regulate a variety of critical gene activity during development (McGinnis et al. 1984; Scott and Weiner 1984). The discovery of homeobox gene conservation across metazoans (multicellular animals) was depicted vividly in several Southern blots of genomic DNA from diverse species (Figure 1). (The Southern blot is a molecular biology technique used to detect the presence of specific DNA sequence fragments by hybridizing a labeled probe that contains a complementary DNA sequence.) It is difficult, in retrospect, to appreciate how surprising this finding was to biologists. What is now a commonplace due to subsequent empirical investigation – "surprisingly deep similarities in the mechanisms underlying developmental processes across a wide range of bilaterally symmetric metazoans . . . a common core of genetic pathways guiding development" (Bier and McGinnis 2003, 25) – was just glimpsed at the time: "some elements of pattern formation in metazoans . . . would seem to be mechanistically related in a very basic way, raising the possibility of universal control mechanisms of development" (McGinnis et al. 1984, 407).

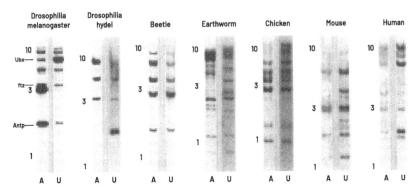

Figure 1 Conservation of the homeobox DNA sequence across metazoans. There are duplicate genomic blots for each species with two different probes containing the ~180 base pair (bp) homeobox sequence: A = 600 bp fragment from the *Antennapedia* homeobox gene of *Drosophila melanogaster* (the fruit fly); U = 450 bp fragment from the *Ultrabithorax* homeobox gene of *Drosophila melanogaster*. Radiolabeled hybridization fragments indicate a complementary DNA sequence and therefore the presence of the homeobox sequence in other species. Ten, three, and one kilobase labels are migration distance size standards. Abbreviations: Ubx: *Ultrabithorax*; ftz: *fushi tarazu*; Antp: *Antennapedia*.
Adapted from: McGinnis et al. (1984). Reproduced with permission from Elsevier.

For many biologists, this memorable event is recalled as a fountainhead: "present-day evo-devo erupted out of the discovery of the homeobox in the early 1980s" (Arthur 2002, 757). "Evo-devo began in the pre-genomic era when genetic studies . . . revealed that the *Hox* genes that control the anterior-posterior (A-P) axis were unexpectedly conserved" (De Robertis 2008, 186). The future, as a consequence, was bright: "We are at a remarkable point in our understanding of nature, for a synthesis of developmental genetics with evolutionary biology may transform our appreciation of the mechanisms underlying evolutionary change and animal diversity" (Gilbert 1997, 914). On analogy with the American Revolution, we might call the visual evidence of homeobox conservation in Figure 1 "the blot seen round the world" (with apologies to Emerson). Yet a more accurate understanding of the emergence and significance of evo-devo in all its complexity turns out to be more complicated (Love 2007b, 2015a; Moczek et al. 2015). And this, in turn, is relevant for how we think about conceptual issues like the structure of evolutionary theory (see Section 1.3).

In 1978, years before the discovery of homeobox conservation, a promising graduate student wrote a letter to his advisor describing his recent intellectual interactions with other biologists at a conference.

> The Gordon Conference on Theoretical Biology was very interesting since I had the opportunity to meet a lot of people in a field that is new for me. The most important event was to meet Lewis Wolpert. He was very interested in our paper and we had a long discussion about the role of development in evolution. He also believes that "the next major breakthrough in biology will involve the integration of development in evolutionary theory." (Pere Alberch to David Wake, July 8, 1978; courtesy of David Wake)

Alberch's conversation with Wolpert became the impetus for a workshop on evolution and development held in 1981 with participants drawn from a variety of disciplinary approaches (e.g., mathematical biology, paleontology, morphology, molecular biology, evolutionary genetics, developmental genetics, and experimental embryology) and taxonomic specialties (lower eukaryotes, marine invertebrates, terrestrial arthropods, and vertebrates). It included Stephen Jay Gould whose historical treatment of the relationship between ontogeny and phylogeny contributed to the motivation behind the workshop (Gould 1977). The resulting edited volume (Bonner 1982) was catalytic for evo-devo and was quickly joined by a chorus of other books (e.g., Arthur 1984; Goodwin et al. 1983; Raff and Kaufman 1983).

The goal of this "Dahlem workshop" on evolution and development was "to examine how changes in the course of development can alter the course of evolution and to examine how evolutionary processes mold development." This remains an apt statement of the two main axes within contemporary evo-devo: (1) *the evolution of development*, or inquiry into the patterns and processes of how ontogeny (development) varies and changes over time; (2) the *developmental basis of evolution*, or inquiry into the causal impact of ontogenetic processes on evolutionary trajectories, both in terms of constraint and facilitation (Love 2015b). However, this description leaves obscure the major role of phylogenetic reconstruction that had not yet matured at the time of the Dahlem workshop (Love 2015a). Both the cladistic revolution in systematics and the increasing use of molecular data to build phylogenetic trees led to a dramatic reconceptualization of the metazoan tree of life (Adoutte et al. 2000). This began with the use of ribosomal RNA to reconstruct relationships among ten different phyla (Field et al. 1988) and fostered the resolution of new major clades, such as the Lophotrochozoa or "crest/wheel" animals (Halanych et al. 1995) and Ecdysozoa or molting animals (Aguinaldo et al. 1997). Additional refinements continue, but the broad outline is now consensus (Figure 2; see Giribet and Edgecombe 2020). Phylogeny and cladistics are recognized as central to research methodology in evo-devo (Jenner 2000; Telford and Budd 2003).

Combining the historical significance of both evolutionary conservation in genetic mechanisms that control development and molecular phylogenetic

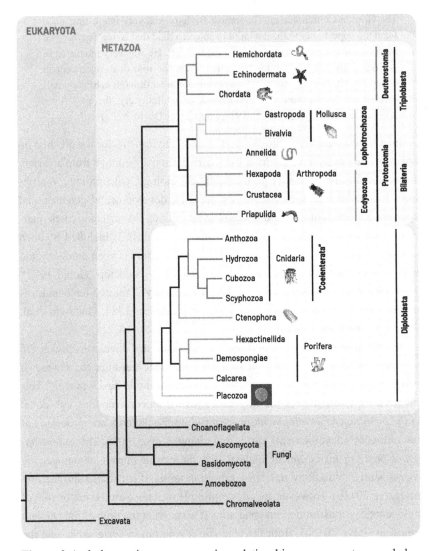

Figure 2 A phylogenetic tree representing relationships among metazoan clades. The reconstruction was done using maximum likelihood methods. Redrawn.

Source: Schierwater et al. (2009). https://commons.wikimedia.org/wiki/File:Metazoan_Phylogenetic_Tree.png.

reconstruction clarifies how crucial elements of contemporary evo-devo emerged. However, these tend to paint a predominantly "genetic" portrait (Wilkins 2002). This is how some biologists interpreted the situation. "Three main factors have contributed to the emergence and phenomenal growth of evolutionary developmental biology. Ironically, all three depend on genetics" (Holland 1999, C41). These three factors are conserved regulatory genes that play similar functional roles in ontogeny across widely divergent taxa, molecular phylogenetics, and molecular biological advances in technique that facilitate the sophisticated analysis and manipulation of genetic material. The first of these yielded the influential concept of a "conserved mechanism" (Section 3) that helped underwrite the increasingly common use of model organisms (Section 2). Arguably, the centrality of the homeobox story in historical memory owes a lot to the centrality of conserved mechanisms and model organisms in everyday experimental research four decades later.

However, keeping in mind the heterogeneity of researchers at the Dahlem workshop, this portrait – focused on molecular advances and genetic manipulation – can encourage a neglect of the interdisciplinary nature of evo-devo (Section 5), which was evident prior to these key developments and provided the intellectual context for the significance of the three factors. For example, many of the contemporary tools for exploring the evolution of development arose from the lineage of experimental embryology, most notably those now ubiquitous in developmental genetics. The dominance of genetic techniques in contemporary developmental biology reinforces their importance to the history of evo-devo (Fraser and Harland 2000). If we shift our attention away from the investigative tools to the agenda of problems within evo-devo, experimental embryology is not the appropriate intellectual ancestor. This agenda, which included reconstructing phylogenetic relationships among metazoan groups and accounting for the origin of evolutionary novelties, derived from comparative evolutionary embryology (Love and Raff 2003), as well as research in morphology and paleontology (Love 2007b).

This historical perspective should, at a minimum, alert us to be cautious about the intended referent when someone refers to evo-devo. Narrow depictions often revolve around the comparative developmental genetics of metazoans (Carroll 2005a; De Robertis 2008), where the focus is on conserved genetic regulatory networks (GRNs)[1] and signaling pathways underlying developmental processes (commonly referred to as "the genetic toolkit"). Evolutionary change is understood in terms of processes of gene regulation with a special emphasis on *cis*-regulatory elements[2] (Carroll 2008; Davidson 2006; Davidson

---

[1] A GRN is an organized collection of molecular entities that interact with one another in coordinated patterns (e.g., negative feedback relationships) to regulate gene expression during development and throughout the life of an organism (Davidson and Peter 2015).

[2] A *cis*-regulatory element is a region of DNA that does not code for a protein but regulates nearby genes through the activation or repression of their transcription derived from transcription factors binding at sites in this region. The adjective "cis" derives from the Latin prefix meaning "on this side" and refers to being located on the same stretch of DNA being regulated (see Davidson and Peter 2015).

and Peter 2015). Most of this empirical research has been prosecuted using model organisms from mainstream developmental biology, such as fruit flies, because the experimental tools available for manipulating these systems are the most powerful and diverse (see Section 2).

Broad depictions of evo-devo include comparative developmental genetics but also point to comparative embryology and morphology, experimental investigations of epigenetic dynamics (e.g., interactions among cell collectives), computational or simulation-oriented inquiry, paleontology, and phylogeny (Hall 2002; Müller 2007; Raff 2000, 2007; Telford and Budd 2003; Wagner et al. 2000). These depictions are sometimes articulated in terms of disciplinary contributors or methodological approaches: "[Evo-devo] is not merely a fusion of the fields of developmental and evolutionary biology, ... [it] strives to forge a unification of genomic, developmental, organismal, population, and natural selection approaches to evolutionary change. It draws from development, evolution, paleaeontology, molecular and systematic biology, but has its own set of questions, approaches and methods" (Hall 1999, xv). These broader depictions have a strong resonance with the more complicated architecture identified from our brief romp through recent history and is reinforced by other analyses (Amundson 2005; Laubichler and Maienschein 2007).[3]

Answers to the question of what evo-devo is vary. And this is not something artificially concocted by historians or philosophers looking in on the field from the outside. "What is evo-devo? This is a hard question to answer succinctly; evo-devo is developmental genetics, it is mathematics and modeling, it is genomics and anatomy and paleontology. In short, evo-devo insights can come from myriad lines of research. This is simultaneously one of the most exciting attributes of the field and one of its greatest challenges" (Albertson 2018, 191). From a philosophical perspective, different answers to the question involve commitments about the structure of knowledge and especially hierarchical relations between different elements (i.e., whether one domain is more fundamental than another). Thus, our history brings us to the present because these commitments are what is at stake in current conversations about evo-devo and calls for an "extended evolutionary synthesis" (EES) (Futuyma 2017; Laland et al. 2015; Müller 2017; Pigliucci 2007; Pigliucci and Müller 2010). One strategy for exploring these commitments is to ask a related question: Within the body of knowledge referred to as evolutionary theory, where do you put evo-devo (Fábregas-Tejeda and Vergara-Silva 2018)?

---

[3] A comprehensive treatment of the landscape of modern evo-devo that nicely covers the diversity described here in much more detail (historical, philosophical, and scientific) can be found in *Evolutionary Developmental Biology: A Reference Guide* (Nuño de la Rosa and Müller 2021). Several of its chapters are cited herein.

1.2 Situating Evo-Devo in Evolutionary Theory

The place of evo-devo within evolutionary theory is contested. Some voices have been skeptical: "problems concerned with the orderly development of the individual are unrelated to those of the evolution of organisms through time" (Wallace 1986, 149). Others have been optimistic, seeing evo-devo as providing a unified theoretical framework in terms of central organizing mechanisms, such as GRNs (Laubichler 2009), or concepts like evolvability (Hendrikse et al. 2007; see Section 4.4). These perspectives are intimately connected to how both evo-devo and standard evolutionary theory are conceptualized: "[evo-devo] seeks to amplify and extend the modern synthesis of evolutionary biology and genetics to include developmental genetics as well as population genetics" (Gilbert and Burian 2003, 68). They appear more specifically in the context of problem domains, such as the origin of evolutionary novelty: "It does not help much to say that there were one or two mutations that created eyespots and that these alleles were selected" (Wagner 2000, 97).

One way of situating evo-devo given this line of criticism is by distinguishing two explanatory projects within evolutionary biology. The first is associated with a perspective that emphasizes populations and function. Evolutionary change from one phenotype or trait to another is explained via population processes such as natural selection, which sorts phenotypes, alters gene frequencies, and yields outcomes that are frequently though not exclusively adaptive (Dickins and Rahman 2012; Hoekstra and Coyne 2007; Lynch 2007). The second project is associated with organisms and structure (Wagner 2014). Evolutionary change from one ontogeny to another is explained by a variety of developmental genetic and epigenetic processes, which can be altered in different ways to produce novel morphologies in organisms (Amundson 2005; Calcott 2009; Laubichler 2010). "Evo-devo represents a causal-mechanistic approach towards the understanding of phenotypic change in evolution. . . . it seeks to explain phenotypic change through the alterations in developmental mechanisms" (Müller 2007, 945–6; see Baedke 2021).

Although this interpretive possibility bolsters the rationale for distinguishing evo-devo from the evolutionary inquiry seen in population genetics or behavioral ecology, it does little for situating evo-devo within evolutionary theory. For example, is the organism-structure explanatory project subsidiary to the population-function explanatory agenda, a prerequisite for it, or simply orthogonal? Some evolutionary biologists have a clear answer to this question: "The litmus test for any evolutionary hypothesis must be its consistency with fundamental population genetic principles . . . population genetics provides an essential framework for understanding how evolution occurs" (Lynch 2007, 8598).

As might be expected, some evo-devo proponents disagree: "I am not con-
vinced that what we have learned about the evolution of form is being
adequately considered in comparative genomics and population genetics,
where the potential role of regulatory sequence evolution appears to be
a secondary consideration, or ignored altogether" (Carroll 2005b, 1164).

Sean Carroll's accent on the "the evolution of form" is an important signal of
where evo-devo might be located within the larger palette of evolutionary
biology because the need to explain the nature and origin of morphology
seems to require a substantive incorporation of embryology and molecular
developmental biology.

> Evolutionary developmental biology (evo–devo) emerged as a distinct field
> of research in the early 1980s to address the profound neglect of development
> in the standard modern synthesis framework of evolutionary theory,
> a deficiency that had caused difficulties in explaining the origins of organis-
> mal form in mechanistic terms. (Müller 2007, 943)

> At the time of the "Modern Synthesis" of evolutionary theory that drew together
> various disciplines including genetics, paleontology, and systematics, very little
> could be said about the effects of genes on development, let alone on the
> evolution of form. . . . [Evo-devo] discoveries forced . . . evolutionary biologists
> to confront a new source of unforeseen and penetrating genetic insights into the
> generation and diversification of animal form. (Carroll 2008, 25)

The referent of "form" in these claims is structure or morphology – the
composition and arrangement, shape, or appearance of organic materials. The
contrast is with function – activities performed or displayed by organisms.

A different yet complementary way of situating evo-devo in relation to evolutionary theory is to see it as addressing the *introduction* of variation, whereas
standard evolutionary models have treated the *fate* of variation (e.g., Stoltzfus
2021). Instead of two different explanatory projects, the evolutionary process is
split into two distinct components. This permits discussion of whether properties of
mutational processes or development that channel variation can bias the rate or
direction of evolution with lesser, similar, or greater strength than processes
involved in the fate of that variation, such as natural selection (Arthur 2004).
Traditionally, this was discussed under the rubric of developmental constraints
that retard natural selection (Maynard Smith et al. 1985). In recent decades,
discussion has shifted to an emphasis on the properties of lineages that contribute
to evolvability (Brigandt 2015a; Stoltzfus 2021; see Section 4.4). What is at stake
can be illustrated concretely in a classic embryological experiment that looked
at the order of condensation formation in amphibian digit development to
account for evolutionary patterns of digital reduction in lineages of this clade

(Alberch and Gale 1985). Frogs experience the loss of preaxial digits ("big toes") during hind limb digital reduction because these digits form last during ontogeny; salamanders, in contrast, lose postaxial digits ("pinky toes") during hind limb digital reduction because these digits are formed last during ontogeny in this lineage (Figure 3). The developmental pattern of how variation is introduced in these lineages helps to explain the evolutionary outcome in a way that selection for hind limb reduction (i.e., the fate of digital variation), which is presumed to have been operative, does not.

These possibilities for how to think about situating evo-devo in evolutionary theory are by no means exhaustive (Love 2020). These and other strategies have merit but depend crucially on how evo-devo is characterized, and there are several distinct ways to accomplish this (Section 1.1).[4] Thus, instead of attempting to settle on what might be the best strategy for situating evo-devo in evolutionary theory, it is preferable to explicitly attend to what we think

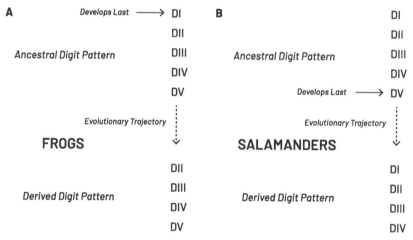

Figure 3 Digital reduction trends in frogs and salamanders. A simplified, schematic representation of how the order of condensation formation in amphibian digit development (the introduction of variation) helps to explain the evolutionary pattern of digital reduction in these two lineages (Alberch and Gale 1985). (A) Frogs experiencing hind limb digital reduction lost preaxial digits ("big toes") because they formed last during ontogeny. (B) Salamanders experiencing hind limb digital reduction lost postaxial digits ("pinky toes") because they formed last during ontogeny. Redrawn.
Source: Love (2015b).

---

[4] This includes questions about the meaning and status of typological thinking in evo-devo (Amundson 2005; Brigandt 2021; Love 2009b).

evo-devo is and the goals of our attempts to situate it. The latter might include sorting out a particular controversy, establishing connections between disciplines, defining the horizons of current research, or clarifying research questions (Scheiner 2010). Some forms of situating evo-devo, characterized in a particular way, may be better suited to achieving some of these goals rather than others. However, these efforts also depend on how we characterize evolutionary theory and therefore it also requires explicit consideration.

1.3 A Plea for Pluralism (About Theory Structure)

An unstated assumption of this discussion about situating evo-devo is that there is a single evolutionary theory in which to locate it. To speak of an extended or expanded theory of evolution suggests we know where the extensions would go or what portions need expansion. This can be seen in the claims educed to contrast standard evolutionary theorizing assumptions from those of an EES (Laland et al. 2015). These include a broadened notion of inheritance beyond genetic material (e.g., cultural and ecological inheritance) that yields an expanded view of what counts as heritable traits of a population and thereby the nature and types of evolutionary processes. This then could account for macroevolutionary patterns and contribute to an understanding of evolvability. Common terminological choices in these descriptions align with a metaphor of spatial increase: "encompass," "broaden," and "additional."

To expand evolutionary theory is to increase its content; to extend it is to enlarge the boundaries such that more things are encompassed. The argument for expansion or extension as the appropriate modifiers for empirical and theoretical developments that EES proponents advance derives from the perceived need to preserve much of what is currently in evolutionary theory. Diagrammatic representations depict clearly that what is in view involves a form of spatial increase (Figure 4). However, we might wonder what the conceptual "chunks" depict in these representations: How do we conceptualize the content of evolutionary theory? These representations do not tell us much about how those chunks might be interrelated, which seems to be a natural assumption if they compose the same theory: How is the content of evolutionary theory organized or structured? At most, the diagram suggests a "core" or foundational structure (center of Figure 4). Often this is thought of as the apparatus of evolutionary genetics that undergirds an understanding of how genotypic and phenotypic changes occur in populations due to natural selection, mutation, migration, and genetic drift.

Ideas about theory content and structure in philosophy of science supply a variety of answers to these questions. For content, we might separate out empirical findings or knowledge about a domain, models of how phenomena in

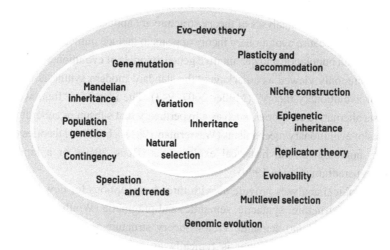

Figure 4 Representing evolutionary theory and the extended evolutionary synthesis (EES). Key evolutionary concepts organized schematically in terms of Darwinism (center field), the Modern Synthesis (intermediate field), and the EES (outer field), representing a trend of continuous expansion. Redrawn. **Source:** Pigliucci and Müller (2010a, figure 1.1, p. 11). © 2010 Massachusetts Institute of Technology, by permission of The MIT Press.

these domains behave, and concepts that represent diverse instantiations of the phenomena that are being predicted, characterized, manipulated, or explained (or doing the explaining). This facilitates discussion of whether new empirical findings must be added, models of phenomena require revision, or new concepts must be invoked to accomplish one or more scientific aims. For structure, there are descriptive questions about how empirical findings, models, and concepts are organized: How is knowledge referred to as "evolutionary theory" ordered and arranged (Bock 2010; Tuomi 1981)? Does it have a structure that is similar to or different from other scientific theories (Lloyd 1988)? There also are prescriptive questions: *Should* we organize evolutionary theory in a particular way? *Should* it be similar in structure to other theories? Questions of this kind draw us into a philosophical conversation about the nature of scientific theories (Winther 2021). These questions are not wholly separated from the content of a theory, but they require exploring evolutionary theory from an angle that is less frequently contemplated by working biologists.

The epistemic organization of evolutionary theory has long worried historians and philosophers of biology (Depew and Weber 1996). It seems anomalous in comparison with other scientific theories, which raises a question about whether existing philosophical analyses of theory structure are applicable (Caplan 1978). In part, this is because evolutionary theory appears to include numerous commitments beyond population and quantitative genetic models of evolutionary change and is commonly viewed as a package: "the standard modern synthesis framework of evolutionary theory" (Müller 2007, 943). The idea of a "framework" suggests alternative categories, such as a hypertheory that subsumes subordinate theories of evolutionary mechanisms (Wasserman 1981), a bundle of theories that involve nomological and historical explanations (Bock 2010), or a kind of Kuhnian paradigm (Gayon 1990).

Philosophical analyses provide us with three basic options for how to think about theory structure: syntactic, semantic, and pragmatic (Winther 2021). In both the syntactic and semantic views, theory structure is reconstructed in a formal language. The former is typically an axiomatization in terms of predicate logic, where the axioms are often understood as laws that, when combined with initial and boundary conditions, will explain the requisite domain of empirical findings.[5] For the semantic view, reconstruction typically occurs in set theory as a family of models that map onto those used in the science. The pragmatic view is more heterogeneous and serves as a broad catchall for approaches to theory structure that hew closely to scientific practices of theory use and therefore include not only mathematical formalism but also a variety of other elements, many of them nonformal (e.g., analogies, assumptions, and heuristic principles). There is no presumption that the contours of theory will be shared across all scientific disciplines and an emphasis on both the purposes to which theory is put and the values of the research communities that use its different elements regularly.

The concerns of biologists about structural organization and arrangement largely suggest a pragmatic orientation, though not because the other two approaches are incorrect. For example, there are fruitful and illuminating analyses of evolutionary theory from the vantage point of the semantic view (Lloyd 1988). However, the pragmatic view keeps our attention on how biologists use evolutionary theory in practice. And this is something observable at our juncture of concern, the relationship of evo-devo to evolutionary theory.

> But do questions posed about evo-devo and evolutionary theory matter to anyone besides the specialists and a few future historians? I think the answers matter very much. By "theory" here, I mean "structures of ideas that explain

---

[5] Although atypical, a few biologists hold views of theory structure amenable with this reconstructive approach: "evolutionary theory is not just a collection of separately constructed models, but is a unified subject in which all of the major results are related to a few basic biological and mathematical principles" (Rice 2004, xiii).

and interpret facts" … Without theories to organize and interpret facts, without the power of general explanations, we are left with just piles of case studies. Moreover, we are without the frameworks that enable us to make predictions about any particular case. (Carroll 2008, 25)

These "structures of ideas" do not come from standard evolutionary theory: "the evolution of form is the main drama of life's story, both as found in the fossil record and in the diversity of living species" (Carroll 2005a, 294). General explanations, in this instance, do not rely primarily on evolutionary genetics or its theoretical principles. Instead of "nothing makes sense in evolution except in the light of population genetics" (Lynch 2007, 8597), we get "nothing in evolution makes sense except in the light of cell, molecular, and developmental biology" (Kirschner 2015, 203). We need more refined tools for theory structure than appeals to expansion or extension can supply.

I suggest we combine the pragmatic orientation to theory structure with a pluralist stance. There is more than one legitimate way to structure evolutionary theory and locate evo-devo in relation to it. When different research questions are in view and different aims are undertaken, divergent explanatory goals and criteria of adequacy come into view and lead to distinct configurations of various theory elements (Love 2010a). But what does it mean to say that evolutionary theory has more than one legitimate structure? What does it mean to adopt a pluralist stance toward theory structure? First, it involves a rejection of the presumption that "there is a single perspicuous representation system within which all correct accounts can be expressed" (Kellert et al. 2006, xv). Second, it takes seriously not only the heterogeneity of theory structure in scientific practice across disciplines but also what we find within disciplines, such as evolutionary biology. There is an embrace of "an openness to the ineliminability of multiplicity in some scientific contexts" (xiii). Instead of engaging in reconstructive efforts that often involve the postulation of "hidden" theory structure that is not present in scientific discourse, we stay focused on how evolutionary biologists themselves create and use theory structures of different kinds to evaluate their own reasoning.

A pluralist stance from a pragmatic orientation concentrates attention on the legitimacy of structures for evolutionary theory in terms of their success or failure with respect to different roles that contribute to scientific aims, such as: (1) articulating the relationships among different scientific disciplines in the theory (or parts of it); (2) facilitating ongoing research with respect to particular parts of a theory, including attempts to answer specific research questions; (3) presenting aspects of the theory in pedagogical contexts for others to assimilate; or (4) comprehending the historical development of models or concepts of the theory. There is no reason to think that success with respect to one or more of these roles will beget success with respect to another. Different roles have distinct and sometimes divergent criteria for deciding when and where different theory structures obtain legitimacy. A pluralist stance not only leads us to expect diverse structures but also helps to explain why they diverge in the ways that they do.

Different theory structures can be interpreted as theory "presentations" (Griesemer 1984), which refer to how empirical findings, models, and concepts are specified in relation to research questions. A presentation of evolutionary theory or some subset of it is legitimate when it produces advantages to some scientific end, such as facilitating ongoing investigation into natural phenomena within a domain. How scientists use evolutionary theory in practice guides preferences about its structural representation; choices are made in terms of what we do with the theory (Love 2013). Theory presentations are always partial specifications that are heuristic in nature and thus prone to systematic biases (Wimsatt 2007). They often appear as summary narratives of the evolutionary process.[6] We can understand them as a kind of modeling that involves idealization – knowingly ignoring variation in properties and excluding values for variables (Jones 2005; Weisberg 2013). Researchers knowingly misrepresent aspects of a theory, whether empirical findings, models, or concepts, in the endeavor of addressing research questions or pursuing other scientific aims.

One fruitful approach to idealized theory presentation is in terms of structured problem agendas (Love 2010a). Within this idealization, the relatively stable topics of evolutionary biology found across textbooks correspond to organized arrays of conceptual, theoretical, and empirical research questions ("problem agendas") about complex phenomena that are concurrently tackled by multiple disciplinary approaches. This idealization offers a way to understand historical continuity in evolutionary theory (e.g., specific empirical questions can change while the problem agenda retains its identity), draws attention to how different disciplinary communities make distinct explanatory contributions (in the context of different problem agendas), displays how research is guided in particular directions (by research questions in a problem agenda rather than hypothesis testing), and accounts for methodological choices and the adoption of explanatory standards. It also displays how different topics in evolutionary theory coalesce around major divisions of biological science – genetics, cell, and developmental biology; ecology; systematics – and explains

---

[6] E.g., "evolution is concerned with inherited changes in populations of organisms over time leading to differences among them. … Genes within individuals (genotypes) in a population, which are passed down from generation to generation, and the features (phenotypes) of individuals in successive generations of organisms do evolve. Accumulation of heritable responses to selection of the phenotype, generation after generation, leads to evolution" (Hall and Hallgrímsson 2008, 3).

why concentrating on one or more problem agendas in evolutionary theory to the exclusion of others can lead to a devaluing of developmental or ecological factors in our accounts of evolutionary processes (Figure 5).

Importantly, this idealized presentation of evolutionary theory does not capture some features we often associate with evolutionary theory. It does a poor job of recovering the distinction between pattern and process. If evolutionary theory is understood in terms of mechanisms of change, then the inclusion of classification is incongruent. However, this is exactly what we would expect from distinct choices that motivate different theory presentations. The idealized picture in Figure 5 makes it difficult to see how phenotypic plasticity (the environmental induction of different traits) and niche construction (organisms actively altering their environment) fit into evolutionary theory, both of which have important places in evo-devo (e.g., Moczek et al. 2011) and discussions of an EES (Laland et al. 2015). Phenotypic plasticity is a phenomenon that combines variation and

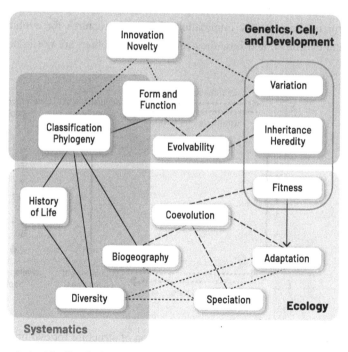

Figure 5 An idealized picture of an erotetic structure for evolutionary theory. This structure is composed of problem agendas and their interrelations, as well as their correspondence to primary domains of biological inquiry: systematics, ecology, and genetics, cell, and development. "Erotetic" means "of or pertaining to questioning" and derives from the Greek word "erōtētikós." Redrawn. **Source:** Love (2010a).

ecology; niche construction is a phenomenon that combines inheritance and ecology. These connections are opaque and difficult to recover on this theory presentation. Yet this was intentionally accomplished; we knowingly misrepresented those aspects. It is an idealization. We can put forward another idealized theory structure (Figure 6) that represents some of these drawbacks (e.g., a focus on evolutionary process, depiction of the role of phenotypic plasticity and niche construction, and the juxtaposing of ecology with variation and inheritance). However, we lose advantages associated with the previous idealized representation (e.g., how different disciplinary communities make distinct contributions or what accounts for methodological choices and the adoption of explanatory standards). Linguistic and pictorial theory presentations always involve trade-offs.

The diversity of theory structures observable in presentations of evolutionary theory across different realms of investigation can be reinterpreted as sets of incompatible idealizations that need not be resolvable into a single perspicuous representation system. We do not have to invoke hypertheories or Kuhnian paradigms to dissect the structure(s) of evolutionary theory or fuss over what is (or is not) in the core. By implication, different structures for evolutionary theory and evo-devo, and how they relate to one another, are not in constant

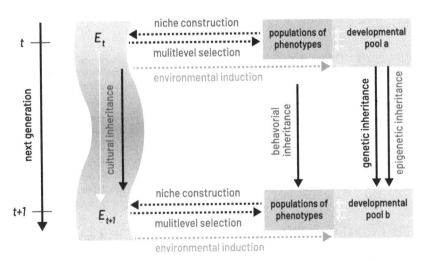

Figure 6 A different idealized representation of structure for evolutionary theory. This structure is composed of elements that emphasize the process of evolutionary change and specifically depict the roles of phenotypic plasticity (via environmental induction) and niche construction, as well as different channels of variation and inheritance. For comparison, see the narrative summary in footnote 6. Redrawn.

Source: Müller (2017).

competition. Whether they are considered rivals depends on whether they share scientific aims. The diversity of methodological and epistemological goals in evolutionary theory implies that a pluralist stance is a good strategy for advancing our understanding of evolution: "the explanatory and investigative aims of science [may] be best achieved by sciences that are pluralistic, even in the long run" (Kellert et al. 2006, ix–x).

2 Model Organisms (and More)

2.1 Criteria for and Categories of Model Organisms

The demonstration of homeobox gene conservation that encapsulated the generative moment recalled by many biologists as launching evo-devo was accomplished in model organisms like the fruit fly, chicken, and mouse (Section 1, Figure 1). Model organisms are central to contemporary biology and studies of embryogenesis in particular (Ankeny and Leonelli 2011; Bier and McGinnis 2003). Biologists utilize only a small number of species to experimentally elucidate various properties of ontogeny. These experimental models permit researchers to investigate development in great depth and facilitate the detailed dissection of causal relationships. A common justification for their use is the conserved genetic mechanisms shared by all metazoans (Kirschner and Gerhart 1998), such as collinear expression of *Hox* genes to specify body axes (McGinnis and Krumlauf 1992). The centrality of conserved mechanisms in model organism research helped to cement the homeobox discovery as the initiating moment for evo-devo (Section 1.1). What exactly a conserved mechanism amounts to will be our focus in Section 3. The present section explores criteria and examples of model organism use in biology, with a special focus on evo-devo where there has been explicit concern about their status (Bolker 2014; Jenner and Wills 2007; Love and Yoshida 2019; Minelli and Baedke 2014).

Developmental model organisms are typically used to establish core similarities exemplified by many taxa, especially with an eye toward medical application (Bier and McGinnis 2003). Critics have questioned whether these models are good representatives of other species because of inherent biases involved in their selection, such as rapid development and short generation time (Bolker 1995), as well as problematic presumptions about the conservation of gene functions and regulatory networks (Lynch 2009). Biologists also routinely discuss "non-model" organisms (Russell et al. 2017). To avoid a game of claim and counterclaim about what counts as a good model (or non-model) organism, it is helpful to focus on two elements of scientific practice: (1) *criteria* that biologists use to evaluate model organisms, and (2) different *categories* of model organisms that have been introduced to capture standardized combinations or specifications of these criteria in different arenas of inquiry.

Two major criteria used to evaluate model organisms are *representation* and *manipulation* (Love and Travisano 2013).[7] The former concerns what biological systems a model can represent and to what extent. The latter concerns ease of empirical examination of a model. Ankeny and Leonelli (2011) emphasize two dimensions of the representational criterion: scope and target. *Representational scope* describes how widely and to what biological systems the results and lessons learned by studying a model system are projected. If a research group uses zebrafish to learn about vertebrates, then zebrafish as a model organism is supposed to have vertebrates as its representational scope. *Representational target* indicates what biological phenomena are explored by studying the model system. If the research group uses zebrafish to explore how neural ectoderm becomes hindbrain, then nervous system development is the representational target. In many cases, the choice of representational target is connected to representational scope. For example, the study of genetic and cellular *mechanisms* in development (representational target) and the discovery that they are widely conserved evolutionarily (representational scope) is a significant motivator for the continued use of model systems (Ankeny and Leonelli 2011). Additionally, the representational target might be a higher-level *phenomenon* instead of a molecular mechanism, and either of these can be scrutinized narrowly (*specificity*) or with respect to a range of variation on the theme (*variety*) (Table 1).

Many different factors are relevant to the criterion of manipulation. Examples include organism availability, the cost of initiating inquiry, possible experimental techniques that can be applied, and how quickly one can produce data and results (Ankeny and Leonelli 2011; Love and Travisano 2013). Although there are differences in the degree to which these factors must be present and how they are fulfilled (e.g., availability might be achieved through chemically preserving and storing specimens), it is widely accepted that manipulation is a crucial criterion for model organisms. Furthermore, the criteria of representation and manipulation are interrelated. In some cases, they exhibit a trade-off relationship; a model that faithfully represents organisms of interest might be difficult to experimentally manipulate, or an organism that is easy to manipulate might represent a phenomenon of interest poorly. Biologists choose and evaluate

---

[7] Dietrich et al. (2020) provide a fine-grained analysis of twenty different criteria for choosing a research organism under five thematic clusters. It is especially detailed with respect to the diversity of pragmatic criteria involved (e.g., ease of supply, viability and durability, training requirements, and institutional support).

Table 1 Representational criteria for model organisms.

Representational scope	*Is related to*	Phenomena or mechanisms	*Focused on*	Specificity or variety
Representational target	*Is*	Phenomena or mechanisms	*With respect to*	Specificity or variety

A summary depiction of the representational criteria for model organisms. Reading from left to right generates phrases that summarize motivations for model organism practices. For example, representational scope can be related to developmental mechanisms, such as collinear *Hox* gene expression, focused on their variety (e.g., establishing anterior–posterior axes of the body versus establishing proximal–distal axes of appendages). Or, the representational target can be developmental phenomena, such as gastrulation, with respect to their variety (e.g., delamination, epiboly, ingression, invagination, and involution).

model organisms by considering these criteria jointly, giving weight to different factors in terms of what they aim to accomplish (i.e., their research purposes).

Configurations of different combinations and specifications of these criteria yield a variety of model organism categories. Ankeny and Leonelli (2011) distinguish *model* organisms from *experimental* organisms. These differ in their representational scope, representational target, manipulative requirements, and purposes of research. "Model organisms" correspond to a limited number of species that have been widely used in biological research, such as the mouse or the fruit fly. Studies of model organisms have various genetic, developmental, physiological, ecological, and evolutionary phenomena as their target, especially those that occur in organisms generally. Results of these studies (e.g., identified mechanisms) are generalized to a wide range of other species. Important manipulative requirements for model organisms (in this sense) are the ability to undertake genetic analysis and successful standardization to minimize confounding variation (e.g., pure lines or strains). They are studied for the purpose of securing "an integrative understanding of intact organisms in terms of their genetics, development, and physiology" (Ankeny and Leonelli 2011, 319).

In contrast, "experimental organisms" are studied to explore specific biological phenomena. They exemplify the Krogh principle: "For a large number of problems there will be some animal of choice or a few such animals on which it can be most conveniently studied" (Krogh 1929, 202). An example is studying fertilization in sea urchins. Experimental organisms are chosen to answer specific questions, which means that each one typically has specific phenomena as its representational target (Green et al. 2018). In the case of experimental organisms, the results of the research are often generalized or extrapolated more narrowly than results derived from model organisms. They are also not as standardized and less suitable for many forms of genetic analysis. However, this is not an inherent drawback because manipulative requirements vary depending on what questions are being addressed. The giant squid axon was strategic for electrophysiological experiments to ascertain neuronal function because its comparatively large size made it possible to reliably apply voltage clamps.

Another helpful distinction separates *exemplary* and *surrogate* models (Bolker 2009). Exemplary models are studied as examples of larger groups of organisms, such as zebrafish as a model of vertebrates. The results acquired by investigating an exemplary model are generalized to a larger group of which the model species is a member. Developmental biology model organisms are exemplary in this sense: "The motivation for their study is not simply to understand how that particular animal develops, but to use it as an example of how all animals develop" (Slack 2006, 61). On the other hand, surrogate models serve as substitutes for specific biological systems. The results derived from them are extrapolated to a specific target. An example is the use of the mouse as a model for human systems or phenomena (Cheon and Orsulic 2011). The inference is from a proxy to a target instead of from an exemplar to other taxa more generally. Different roles played by exemplary models and surrogate models arise from different investigative purposes. For exemplary models, the goal is to identify widely shared biological mechanisms or better understand evolutionary processes. As a result, they are associated with basic research. In contrast, surrogate models are used in applied fields, such as biomedicine or conservation research, to better understand medical or ecological problems and develop potential solutions.

2.2 Model Clades and Life Histories

Although most accounts of the criteria and categories for model organisms have been concerned with organisms of specific *species*, there are also perspectives that focus on differently arrayed biological systems. The notion of a model *taxon* refers to a clade (a group derived from a common ancestor) that is used to investigate diverse questions about genetic, developmental, physiological, ecological, and evolutionary phenomena within a clade and identify generalizations applicable to other clades (Griesemer 2015). Thus, the representational scope of a model taxon is understood as taxa within and beyond the clade, with results ascertained through investigation applying differentially to individual taxa of smaller and larger sizes. Inquiry is organized around interrelated "packages of phenomena" as the representational target – rather than individual phenomena – and these constitute central features of the evolution of the taxon. For example, lungless salamanders have been studied as a model taxon and provided explanatory insights about the evolution of anatomical features (e.g., the tongue or body size) and associated functional systems (e.g., feeding, locomotion, etc.), especially with respect to mechanisms that contribute to these patterns (e.g., miniaturization derived from changes in developmental timing) (Wake and Larson 1987). Another prominent model taxon used to investigate ecological and evolutionary questions is the squamate genus *Anolis*, which is composed of many lizard species with good phylogenetic resolution, while also being regionally delimited and relatively accessible (Sanger 2012).

Griesemer uses the term "export" to refer to the predominant inference made from model taxa, where exportation is distinguished from inferences involving simple generalization. The latter is common in molecular biology and operates by assuming that conspecific individuals (or species members) are instances of the same type. Exportation is based on the idea that taxa are historical individuals and in genealogical relationships with one other. Although discussions of model organisms tend to focus on how particular results acquired by studying them are extrapolated, generalized, or exported, Griesemer also emphasizes additional kinds of payoffs that can be acquired by studying model taxa, such as methodological lessons about how to investigate different taxa.

Another distinct category is a model *life history* (Love and Strathmann 2018). These are temporal sequences that occur within the ontogeny of organisms, characterized in terms of functional and morphological properties, which are used as models of developmental sequences found in other species. Marine invertebrate larvae are a primary example (Love 2009a). Unlike a model taxon, these stages of life history are not unified within a single monophyletic group or clade. Often, studies of marine larvae involve cross-clade comparisons of functional requirements for specific ecological settings that exhibit a broad but disjoint taxonomic distribution (i.e., their representational scope). The representational targets of investigations of model life histories are functional or morphological traits in ontogenetic sequences relevant to questions about developmental, ecological, and evolutionary phenomena. Some resulting generalizations revolve around instantiations of larval forms that exemplify a broader type (e.g., the trochophore of mollusks and annelids), while others pertain to behavioral and ecological patterns, such as feeding versus nonfeeding or planktonic (floating in the open sea) versus benthic (living on the seafloor). Model life histories concentrate attention on problems related to adaptation and phylogeny. They serve to coordinate research by scientists from different

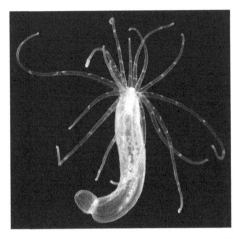

Figure 7 The starlet sea anemone, an evo-devo model system.

Adapted from: Wikimedia Commons (CC BY-2.0). Robert Aguilar, Smithsonian Environmental Research Center. https://commons.wikimedia.org/wiki/File:Nematostella_vectensis_(I1419)_999_(30695685804).jpg

disciplines, often at marine stations where the availability, cost, and infrastructural prerequisites for manipulation are in place to support the relevant configuration of approaches simultaneously.

2.3 Evo-Devo Models: Something Similar, Something Different

For evolutionary researchers, a key investigative aim is establishing significant patterns of difference (i.e., evolutionary change), often manifested across a clade, rather than only sameness (i.e., conservation). However, these two ends of a continuum are entangled in the use of model systems within evo-devo. The establishment of conservation or homology for a character can provide a basis from which to identify and characterize evolutionary divergence in that character or the origin of new ones (see Section 4). Given this entanglement, it is unsurprising that researchers have often reflected methodologically on model organisms in evo-devo (Bolker 2014; Milinkovitch and Tzika 2007; Minelli and Baedke 2014; Sommer 2009), especially in calls to introduce and standardize new ones (e.g., Almudi et al. 2019; Lapraz et al. 2013). "Model organisms . . . are conceptual carry-over from developmental biology, but their study was crucial in establishing evo–devo as a new discipline" (Jenner and Wills 2007, 311). This conceptual carryover includes the criterion of manipulation because experimental tractability is an important aspect of evo-devo model systems. The results acquired by studying experimentally tractable model systems in evo-devo are both generalized or extrapolated to particular taxa (developmental modeling emphasizing similarities) and compared phylogenetically to other taxa (evolutionary modeling emphasizing differences).[8]

The reasoning strategies associated with evo-devo model systems have distinctive features and are not sufficiently characterized by any one of the accounts of model organisms reviewed thus far. One central feature is the importance of phylogenetically informed comparison. Although results and lessons acquired by studying model systems in evo-devo are generalized or extrapolated, as in many other model organisms, this is typically followed by comparisons with taxa related by patterns of common descent. These comparisons are crucial for elucidating the origin of novel traits in a lineage, the evolution of properties of ontogeny, and dissecting the relevant influence of developmental processes on evolutionary mechanisms, all of which motivate the use of model systems in evolutionary developmental research.

A good example of an evo-devo model system is the starlet sea anemone, *Nematostella vectensis* (Figure 7; Darling et al. 2005; Genikhovich and Technau 2009; Layden et al. 2016). It has many practical advantages for experimental studies: easily maintained in little space and at a low cost; adults reproduce under laboratory conditions about once a week and throughout the year; eggs are large enough for manipulation; and the generation time is relatively short. Resources and manipulation techniques available include an annotated genome, *in situ* hybridization and immunohistochemical analysis, and knockdown/knockout techniques from molecular genetic analysis.

An important problem to which *Nematostella* can contribute answers is the evolution of bilaterality. Bilateral symmetry originated in animals more than 600 million years ago, and evo-devo researchers have tried to detect when and explain how the two major body axes (anterior–posterior (A-P) and dorsal–ventral (D-V)) emerged (Erwin 2020). To accomplish this requires comparing bilaterian axis formation (e.g., in *Drosophila* or mouse) with the ancestral pattern of development. *Nematostella* belongs to the phylum Cnidaria, which is an out-group of Bilateria that includes corals, jellyfish, hydras, and sea anemones (Figure 2). It is expected to serve as a model of the ancestral pattern of development in metazoans on the

---

[8] To avoid terminological confusion, I refer to evo-devo model "systems" rather than evo-devo model "organisms" (Love and Yoshida 2019). However, this choice of terminology is for the sake of clarity and does not reflect an underlying, substantive distinction between systems and organisms.

assumption that extant members of this out-group have retained significant features of this pattern. Although it appears to be radially symmetrical, *Nematostella* has two body axes: the oral–aboral axis runs from the mouth to the other end of the body, and the directive axis runs across the pharynx, orthogonal to the oral–aboral axis. Molecular developmental studies have revealed relations of similarity and difference between the body axes of *Nematostella* and the A-P and D-V axes of bilaterians, providing insight into how the primary and secondary axes of symmetry in animals originated.

Consider Wnt/β-catenin signaling, which plays a crucial role in A-P axis specification across bilaterians (Petersen and Reddien 2009).[9] In many bilaterian species, Wnt/β-catenin signaling is differentially activated at certain stages of embryonic development and in different locations of the embryo. During early embryogenesis, the side of the embryo with high Wnt/β-catenin signaling activity develops into the posterior end, while the side with lower signaling activity becomes the anterior end. Because of this, the function of Wnt/β-catenin signaling in oral–aboral axis specification has been examined in *Nematostella*. From the mid-blastula stage where the embryo is a hollow ball of cells, *Wnt* gene expression exhibits a demarcated, staggered expression within the oral half of the embryo (Kusserow et al. 2005). Overactivation of Wnt/β-catenin signaling promotes oral identity, while inhibition leads to expanded expression of aboral markers and reduction of oral marker expression (Röttinger et al. 2012).

These results point toward a role for Wnt/β-catenin signaling in primary axis specification that existed before the separation of Bilateria and Cnidaria (Petersen and Reddien 2009). They also hint at a potential correspondence between bilaterian anterior and cnidarian aboral axes, on the one hand, and between bilaterian posterior and cnidarian oral axes, on the other. However, more recent work demonstrates that the directive axis of *Nematostella* is under the control of an axial *Hox* gene code – a developmental characteristic of the metazoan A-P axis – which indicates that molecular signaling pathways from both axial specification mechanisms in bilaterians are present in directive axis specification (He et al. 2018). This suggests that there is no straightforward relationship of homology between the oral–aboral and directive axes of *Nematostella* and the A-P and D-V axes of bilaterians.

---

[9] The Wnt/β-catenin pathway is made up of four basic segments: an extracellular signal, a membrane receptor, cytoplasmic interactors, and a component that translocates to the nucleus. The extracellular signals are primarily Wnt proteins; a key player in the cytoplasmic interactions and eventual translocation to the nucleus is β-catenin. In development, this pathway initiates gene expression related to cell proliferation and the regulation of cell polarity and migration (van Amerongen and Nusse 2009; see Section 3 for further discussion).

Evo-devo model systems represent a synthesis of generalization or extrapolation of developmental mechanisms and phylogenetic comparison to answer questions about the evolution of development and the developmental basis of evolutionary change. This synthesis exhibits elements of reasoning strategies associated with model organisms, clades, and life histories but ultimately has a distinctive status. Evo-devo model systems are typically one species, which makes them distinct from model taxa and model life histories. To use a model taxon, one studies multiple species in the clade with the aim of exporting the lessons learned to other members of the same clade or to different clades. In contrast, an evo-devo model system is a specific species utilized to produce results that can be generalized or extrapolated to another species rather than across entire clades. For example, increased Wnt/β-catenin signaling in *Nematostella* promotes oral identity in the establishment of the oral–aboral axis (Röttinger et al. 2012), which can be generalized to primary axis specification in a representative ancestral metazoan species extant prior to the split between Bilateria and Cnidaria hundreds of millions of years ago (Petersen and Reddien 2009). By parallel reasoning, evo-devo model systems are distinguished from model life histories because the latter category is applied to specific temporal sequences within development. This does not mean that model life histories cannot help to address evolutionary developmental questions. The crucial point is that they do so in a different fashion, such as by generalizing functional requirements of dispersal or feeding for larval forms during ontogeny in specific ecological settings.

An evo-devo model system can serve as an exemplar of a larger class of species or as a surrogate of a particular species. The former applies when one species belonging to a particular clade (e.g., *Nematostella*) is studied to elucidate how traits characteristic of that clade have evolved (e.g., axial symmetry). Evo-devo model systems also are studied as exemplars of species in a taxon that have been underrepresented in developmental research; information about the developmental mechanisms of species in this taxon is needed to elucidate the evolution of widely distributed (though not necessarily homologous) traits. But how an evo-devo model system exemplifies can change with context. In some cases, *Nematostella* is regarded as an exemplar of anthozoan cnidarians or cnidarians in general, or even as an exemplar of animals that exhibit developmental processes such as asexual fission and regeneration (Burton and Finnerty 2009). In other cases, such as studies related to the evolution of bilaterality, *Nematostella* serves as a surrogate model of extinct metazoans because it has likely retained ancestral features of axis specification and thus could be a proxy for the last common ancestor of Bilateria and Cnidaria. This is reflected in the ways that biologists discuss representational roles in the context of this research: "many ancestral traits have been preserved in *Nematostella* ... this makes *Nematostella* a very attractive model system among the representatives of basal metazoan lineages" (Genikhovich and Technau 2009, 1). Importantly, it does not have to represent "basal" (or ancestral) metazoans with respect to all traits to serve as a surrogate model. What is required in this context is that *Nematostella* represents the last common ancestor of Bilateria and Cnidaria *with respect to body axis specification*. Ascertaining whether it does demands substantial comparative phylogenetic effort (Jenner 2022, chapter 10). Thus, usefulness as a surrogate can vary depending on which trait is under scrutiny.

Although the distinction between exemplary and surrogate models is useful to capture some of the representational roles played by evo-devo model systems, other aspects of this distinction are not readily applicable. Bolker (2009) claims that a major aim of using exemplary models is to elucidate widely conserved mechanisms. However, evo-devo model systems are studied primarily to elucidate developmental mechanisms characteristic of species in a taxon *to compare with and discover changes in* corresponding developmental mechanisms of species in other taxa. Discovering that these mechanisms are widely conserved is not the purpose of investigation – the aim is to uncover how these developmental mechanisms evolved. Additionally, surrogate models are often used to understand disease etiology, identify possible therapies, or conserve threatened species. Although we can find comparative studies of axis specification used to illuminate human vertebral pathology (ten Broek et al. 2012), *Nematostella* is not under comparison in that situation; its use is motivated primarily by an interest in evolutionary questions, such as the origin of bilateral symmetry.

Evo-devo model systems are not model organisms *sensu* Ankeny and Leonelli (2011) because they are not intended to have a wide range of species as the representational scope and diverse phenomena as the representational target. The category of experimental organism appears better suited but also is not entirely appropriate. For example, the relationship between the choice of an evo-devo model system and the research questions being asked is more complex. An experimental organism is a convenient system for studying specific developmental, physiological, genetic, or behavioral phenomenon. Thus, if a species exhibits the phenomenon of interest and satisfies relevant manipulation criteria, it can be a satisfactory experimental organism. Exhibiting specific phenomena is only part of the representational requirement for a species to be an evo-devo model system; its phylogenetic location in an evolutionary tree is also critical. The generalization or extrapolation of experimental results from an evo-devo model system is followed routinely by a comparison between the taxa in which the generalization or extrapolation applies and other taxa where it does not. Such a comparison is crucial to elucidate the evolution of the traits under scrutiny.

The goal of using an evo-devo model system is not to examine mechanisms underlying a particular developmental, physiological, genetic, or behavioral phenomenon, but rather the examination of these mechanisms *as a means to the end* of tackling evolutionary problems. To address how bilateral symmetry originated and evolved, one must answer a variety of associated questions as a prerequisite, such as how the different mechanisms that establish primary and secondary axes in bilaterians operate. How and in what different ways is the A-P or D-V axis determined during bilaterian embryogenesis? If *Nematostella* serves successfully as a surrogate model of the last common ancestor of Bilateria and Cnidaria, then mechanisms of its axis specification can be extrapolated to the ancestral metazoan. Researchers can then compare the (hypothetical, extinct) ancestral patterns of axis specification with those in extant bilaterian models. This comparison is a key step to help account for the evolution of bilaterality.

In summary, evo-devo model systems are chosen and evaluated based on their potential contributions to answering research questions about the evolution of development and developmental basis of evolutionary change. Answering such questions involves comparisons of developmental patterns and mechanisms found in different lineages. Therefore, the precise location of a species within the evolutionary tree is a critical factor. Consequently, evo-devo model systems and experimental organisms are distinct. Unlike experimental organisms, evo-devo model systems are not chosen simply because they are experimentally tractable and exhibit interesting biological phenomena. They have to occupy appropriate phylogenetic positions, as well as exhibiting particular phenomena, so that effective comparisons can be made to answer important questions that comprise the research problems of evo-devo, such as the properties underlying the capacity to evolve (i.e., evolvability) or the origin of novel traits (Jenner 2006; Jenner and Wills 2007; Milinkovitch and Tzika 2007; Minelli and Baedke 2014).

Evo-devo model systems instantiate a distinctive synthesis of model organism approaches from developmental biology and comparative approaches from evolutionary biology. They are experimentally tractable species that act as exemplars or surrogates in that results acquired by studying them can be extrapolated to a specific species or generalized to a larger group. Evo-devo model systems depend heavily on what phylogenetic comparisons they make possible. The purpose of studying them is to answer questions about the evolution of development and the developmental basis of evolutionary change. Strategic comparisons between taxa are an essential step in this methodology, which means the "judicious choice of new model organisms is necessary" (Jenner and Wills 2007, 311). Researchers must navigate the continuum between developmental conservation and evolutionary change by uniquely integrating model organism and comparative approaches to study complex phenomena at the intersection of development and evolution.[10] Answering central questions in evo-devo requires both intensive experimental examination of developmental mechanisms in selected species and phylogenetic comparison of different species within and across taxa. The involvement of phylogenetic comparison as an essential part of evo-devo research makes these model systems a distinctive category that deserves special methodological consideration.

2.4 Evo-Devo Models: The Epistemic Import of Practicalities

Before leaving the topic of model systems in evo-devo, it is worthwhile to examine how practical decisions related to their use have epistemic import. This can be observed in how "normal development" is conceptualized through reasoning strategies that manage variation inherent within and across developing organisms (Lowe 2015, 2016). One strategy is to establish a set of normal stages that break down a complex developmental trajectory into distinct temporal periods by reference to the occurrence of major events, such as fertilization, gastrulation, or metamorphosis (Minelli 2003, chapter 4). This enables researchers in different laboratory contexts to have standardized comparisons of experimental results (Hopwood 2007). However, this strategy ignores developmental variation associated with environmental variables (Gilbert and Epel 2009), such as phenotypic plasticity (the capacity of a particular genotype to generate phenotypic variation), which has been of interest to evo-devo biologists for questions like the origin of evolutionary novelties (Moczek et al. 2011; West-Eberhard 2003). What happens when well-established model organisms are used to experimentally explore the evolutionary significance of phenomena like phenotypic plasticity?

In order to evaluate these questions experimentally, biologists need to alter development through the manipulation of environmental variables and observe the manifestation of variation or how a novel phenotype can be established within the existing plasticity of an organism (Kirschner and Gerhart 2005, chapter 5). This manipulation can identify patterns of variation through the reliable replication of specific experimental alterations within different environmental regimes. However, without measuring variation

---

[10] Although I have focused on animal models, my analysis also applies to plant evo-devo (e.g., Di Stilio and Ickert-Bond 2021).

across environmental regimes, evo-devo biologists cannot detect phenotypic plasticity. These measurements are required to document the degree of plasticity and its patterns for a particular trait, such as qualitatively distinct morphs (e.g., herbivorous versus carnivorous larvae in spadefoot toads; Levis and Pfennig 2019). An evaluation of the significance of phenotypic plasticity for evolution requires answers to questions about where plasticity emerges, how molecular genetic mechanisms are involved, and what genotype–phenotype relations obtain.

Normal stages intentionally ignore variation associated with phenotypic plasticity. Organisms are raised under stable environmental conditions so that stages can be reproduced in different laboratory settings and variation is often viewed as noise that must be reduced or eliminated to understand how development works (Frankino and Raff 2004). This practice also encourages the selection of model organisms that exhibit less plasticity (Bolker 1995). The laboratory domestication of a model organism may also reduce the amount or type of observable phenotypic variation, though it also can increase variation (e.g., via inbreeding). Despite attempts to reduce variation by controlling environmental factors, some of it always remains (Lowe 2015). Evidence of this variation is visible in the fact that absolute chronology is not a reliable measure of time in ontogeny, and neither is the initiation or completion of its different parts (Mabee et al. 2000). Developmental stages allow this recalcitrant variation to be effectively ignored by judgments of embryonic typicality. Normal stages also involve assumptions about the causal connections between different processes across sequences of stages (Minelli 2003, chapter 4). Once these stages have been constructed, it is possible to use them as a visual standard against which to recognize and describe variation as a departure from the norm (DiTeresi 2010; Lowe 2016). But, more typically, variation ignored in the construction of these stages is also ignored in the routine consultation of the stages in day-to-day research contexts (Frankino and Raff 2004).

Normal stages fulfill a number of goals related to descriptive and explanatory endeavors that biologists engage in (Kimmel et al. 1995). They yield a way to measure experimental replication, enable consistent and unambiguous communication among researchers, facilitate accurate predictions of developmental phenomena, and aid in making comparisons or generalizations across species. They allow for a classification of developmental events that is comprehensive with suitably sized and relatively homogeneous stages, reasonably sharp boundaries between stages, and stability under different investigative conditions, which encourages more precise explanations within disciplinary approaches (Dupré 2001). However, these advantages are accompanied by several drawbacks. Key morphological indicators sometimes overlap stages, terminology that is useful for one purpose may be misleading for another (e.g., cross-species comparisons), and manipulation of the embryo for continued observation can have a causal impact on ontogeny. Avoiding variability in stage indicators can encourage overlooking the significance of this variation, or at least provide a reason to favor its minimization. In addition to these drawbacks, most model organisms are poorly suited to inform us about how environmental effects modulate or combine with genetic or other factors in development, making it difficult to elucidate mechanisms underlying reaction norms (the range of phenotypes expressed by a genotype along an environmental gradient, such as temperature). Short generation times and rapid development are tightly correlated with insensitivity to environmental conditions (Bolker 1995).

Should evo-devo biologists *not* use model organisms whose development has been periodized with normal stages? This would mean giving up the tremendous experimental leverage these models afford. The documentation of patterns of variation is required to gage the evolutionary significance of phenotypic plasticity. Practices of developmental staging in model organisms can retard our ability to make either a positive or negative assessment. For example, if evolutionary investigations revolve around a character that was assessed for typicality to underwrite the temporal partitions of stages, then much of the variation in this character was conceptually removed when rendering the model organism experimentally tractable.[11] Is there a way to address the tension between the practice of developmental staging and uncovering the relevance of variation due to phenotypic plasticity for evolution?

An affirmative answer can be formulated out of reasoning practices I call "compensatory tactics" that constitute a principled approach to addressing this tension and promote observations of variation due to phenotypic plasticity that is ignored by developmental stages (Love 2010b): the employment of diverse models and the adoption of alternate periodizations. For the former, variation is often observable in nonstandard models because experimental organisms that do not have large communities built around them are less likely to have had their embryonic development formally staged, and thus the effects of strategies of abstraction on phenotypic plasticity are not as operative (a good example is horn morphology in dung beetles; Rohner et al. 2020). In turn, researchers are sensitized to the ways these kinds of variation are being muted in the study of standard model organisms. Stages can be used actively as visual

---

[11] The standard periodization for postembryonic ontogeny in arthropods is a barrier to evolutionary analyses of molt-timing evolution because the conventional periodization is in terms of molt-to-molt intervals after hatching, which are then grouped into stages (larva, pupa, and imago for insects), irrespective of differences in the timing of the molts (Minelli et al. 2006).

standards to identify variation as deviations from a norm and thereby characterize patterns of variability.

The second compensatory tactic – alternative periodizations – involves choosing different characters to construct new temporal partitions, thereby facilitating the observation of variation with respect to characteristics previously stabilized in normal stages. Alternative periodizations often divide a subset of developmental events according to different processes or landmarks. As a result, they usually do not map one–one onto existing normal stages. This lack of isomorphism between normal stages and alternative periodizations also will be manifested if different measures of time are utilized, whether sequence (event ordering) or duration (succession of defined intervals), and whether sequences or durations are measured relative to one another or against an external standard, such as absolute chronology (Reiss 2003). These incompatibilities prevent assimilating the alternative periodizations into a single, overarching staging scheme. Since alternative periodizations require choosing different characters to stabilize and typify when defining their temporal partitions, different kinds of variation will be exposed than were previously observable.[12] Therefore, specific kinds of variation in developmental features related to environmental variables that might be relevant to evolution (i.e., a particular kind of developmental basis of evolutionary change) can be studied systematically by evo-devo biologists even in standard model organisms.

3 Conserved Mechanisms: The Very Idea

3.1 Prolegomenon to a Puzzle

Conserved developmental mechanisms are central to evo-devo. In the previous section, we saw how model organisms that exhibit them play a key role in investigation. We also observed how the conservation of homeobox genes was a crucial element in the growth of contemporary evo-devo and continues to stand out prominently in the memory of biologists (Section 1).[13] However, a conceptual puzzle is embedded within this common category. Mechanisms are individuated by the outcomes they produce (Glennan et al. 2022). Since the claim of conservation is a judgment of homology, which is typically based on

---

[12] To explore wing developmental evolution, Reed et al. (2007) constructed stages for final-instar wing disk ontogeny in the common buckeye butterfly because the timing of larval events is relatively dissociated from wing disk development. Thus, temporal measures of larval ontogeny do not correlate tightly with the developmental state of the wing disks. This facilitated studying phenotypic plasticity: "Some cryptic variation, however, might manifest developmentally or physiologically, but simply not have an effect on phenotypes that is obvious or accessible to investigators" (Reed et al. 2007, 2).

[13] "If you were to ask me, 'what was the single most important discovery in the origin of evo-devo?' I'd reply, with little hesitation, 'the homeobox'" (Arthur 2021, 6).

structure rather than function (e.g., mole forelimbs are for digging and seal forelimbs are for swimming, but they are homologous structures), what constitutes the individuation conditions for a conserved (homologous) mechanism? A mechanism is individuated functionally, but homologies are typically individuated structurally.

Before proceeding, we need some background on the concept of homology plus basic vocabulary and a mechanism example. Richard Owen defined homology as "the same organ in different animals under every variety of form and function" (Owen 1843, 379). "Organ" is indicative of a structure (an entity, like a forelimb) that may vary in its shape and composition (form: claws versus a flipper) or what it is for (function: digging versus swimming) but is found to correspond across organisms where it occurs (Brigandt 2002). Translated into an evolutionary context, the corresponding *sameness* is cashed out by reference to common ancestry. Since structures can also be *similar* by virtue of natural selection operating in similar environments, homology is often contrasted with analogy. Homologous structures are the same by virtue of descent from a common ancestor, regardless of what functions these structures perform, whereas analogous structures are similar by virtue of selection favoring comparable functional outcomes, regardless of common descent ("convergent evolution"). Since structures can also be similar due to shared developmental pathways ("parallel evolution"), the label "homoplasy" is routinely used to capture all nonhomologous trait similarities. However, the conceptual terrain around homology is complicated (see Novick (2018) and Gouvêa (2020) for helpful discussions). For present purposes, we only need to keep in view that homologous structures like forelimbs typically do not have conserved functions.

Now some vocabulary about mechanisms. Philosophical explorations of mechanisms and mechanistic explanation have grown dramatically over the past two decades (Craver and Darden 2013; Illari and Williamson 2012). Although different accounts of mechanisms have been offered, four shared elements are discernible (Craver and Tabery 2016; Glennan et al. 2022): what a mechanism is for, its constituents, its organization, and the spatiotemporal context of its operation. From these we can generate an abstract, ecumenical characterization to guide our analysis of concrete descriptions of mechanisms under scrutiny in many evo-devo studies.

A mechanism is constituted by a number of parts and activities or component operations that are organized into patterns of interacting relationships within a particular spatiotemporal context so as to produce a specific behavior or phenomenon (or set thereof).

Mechanistic explanation involves decomposing systems into constituent parts, localizing their characteristic activities, and articulating how they are organized to produce a particular effect at or within a specific time or place. These explanations illustrate and display the generation of phenomena by describing the organization of a system's constituent components and activities.[14]

Next, an example. Consider the initial formation of segments in *Drosophila* due to the segment polarity network of gene expression (Damen 2007; von Dassow and Odell 2002). About three hours after fertilization, *Drosophila* embryos have fourteen transient parasegment units. The transcription factor Engrailed accumulates in the anterior portion of each one, which initiates gene activity that defines the boundaries of cell compartments that will eventually become segments (i.e., units of the adult body). One element of this activity is the expression of Hedgehog, a secreted signaling protein, in cells anterior to the band of cells expressing Engrailed, which marks the posterior boundary of each nascent segment. This, in turn, activates Wingless, another secreted signaling protein, which maintains the expression of both *engrailed* and *hedgehog* in a feedback loop so that segment boundaries persist (Figure 8).

This mechanism description can be expanded to illustrate interactions in the feedback loop. On one side, the Hedgehog signaling pathway is activated when Hedgehog binds to the membrane protein Patched (Lum and Beachy 2004). In the absence of Hedgehog, Patched inhibits another membrane protein (Smoothened). Once Hedgehog binds to Patched, Smoothened can block the production and operation of repressors of the transcription factor Cubitus interruptus, which then turns on *wingless*. On the other side, Wingless (a Wnt protein family member) jointly binds membrane proteins, which disrupts a complex of proteins in the cytoplasm that continually degrade β-catenin. The phosphoprotein Dishevelled is also activated, further blocking this complex from operating by anchoring it to the plasma membrane. β-catenin then accumulates and reaches the nucleus in sufficient concentrations to initiate transcription and expression of *engrailed* and *hedgehog*. The entire mechanism is labeled a network due to its many complex interactions (Figure 9).

The segment polarity network is extremely complicated with bizarre protein names that often originated from the mutant phenotype in *Drosophila* where they were discovered (e.g., the *hedgehog* mutant embryo is covered in pointy denticles). However, what is important is that it exhibits the features of our ecumenical

---

[14] Nothing here presumes that mechanistic explanations corresponding to this ecumenical form are the primary or only needed approach to explanation in evo-devo. Mechanistic and mathematical models involving entities and activities that are both qualitatively and quantitatively characterized in complex, dynamic networks of feedback and interaction (i.e., organization) all play a role (Baedke 2021; Brigandt 2015b).

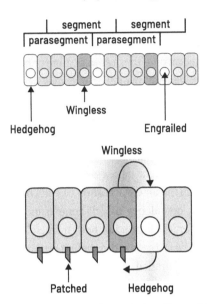

Figure 8 Wingless and Hedgehog reciprocal signaling of the segment polarity network. See text for details.

Adapted from: Wikimedia (CC BY-SA 4.0.). Fred the Oyster. https://commons.wikimedia .org/wiki/File:Wingless_and_Hedgehog_reciprocal_signaling_during_segmentation_ of_Drosophila_embryos.svg.

characterization. It is constituted by a number of parts (e.g., Engrailed, Wingless, Hedgehog, etc.) and activities or component operations (e.g., signaling proteins bind receptors, transcription factors bind to DNA and initiate gene expression, etc.), which are organized into patterns of interacting relationships (the positive feedback loop, the Wnt and Hedgehog signaling pathways, etc.) within a spatiotemporal context (in parasegments of the *Drosophila* embryo, ~3 hours postfertilization), so as to produce a specific phenomenon (a set of distinct segments with well-defined boundaries). The Wnt and Hedgehog signaling pathways are also distinct mechanisms (Lum and Beachy 2004; van Amerongen and Nusse 2009). However, whereas these mechanisms are stable and present throughout the organism's lifetime, the segment polarity network is transient and operative for only a specific period of ontogeny because *hedgehog* and *wingless* become decoupled later in embryogenesis. There is a change in the organization of the mechanism as the spatiotemporal context changes.

3.2 The Conceptual Puzzle (and a Solution)

Given this prolegomenon, we can now articulate the puzzle. Recall our theoretical framing. The claim of conservation for a mechanism is a judgment of

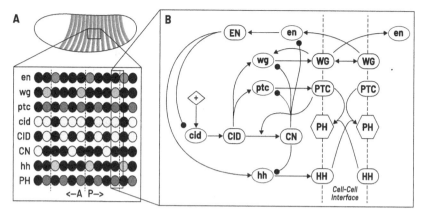

Figure 9 Segment polarity network details. (A) Cartoon representation of a *Drosophila* embryo with an inset of gene expression patterns in cells of the parasegments. Boundaries between parasegments (dashed vertical lines) are defined by the expression of *wingless* (wg) to the anterior and *engrailed* (en) to the posterior (see Figure 3.1 for more detail). (B) Inset depicting a cell–cell interface at the parasegment boundary, which shows the complex nature of interactions occurring in the basic segment polarity network. Abbreviations: CID/cid: cubitus interruptus (whole protein); CN: cubitus interruptus (*N*-terminal repressor); EN/en: engrailed; HH/hh: hedgehog; PH: patched-hedgehog complex; PTC/ptc: patched; Wg/wg: wingless. Redrawn.

Source: von Dassow and Odell (2002). Reproduced by permission of John Wiley and Sons.

homology; the core idea behind homology is identifying the same trait in different taxa under every variety of form and function; sameness derives from a common evolutionary heritage. Similarity is neither necessary nor sufficient for determining whether two traits are homologous (Ghiselin 2005). Many morphological features are similar due to natural selection (i.e., analogous) or shared developmental resources (i.e., parallelism) but not because of common descent (i.e., homologous). How should the sameness of a mechanism that is pertinent to making a homology judgment be conceptualized?

Recall one of the clauses in our ecumenical characterization: "to produce a specific behavior or phenomenon (or set thereof)." This is a reminder that the individuation of mechanisms is *functional*, not structural: "The boundaries of a mechanism . . . are fixed by reference to the phenomenon that the mechanism explains. The components in a mechanism are components in virtue of being relevant to the phenomenon" (Craver and Tabery 2016). A specific behavior or phenomenon – the function of the mechanism – is conserved (i.e., the same) if it does not vary in particular ways. However, a homologue is the same under every

variety of form and *function*. Similarity of function is a problematic criterion of
homology; what a trait does should not be the basis of an evaluation of
homologue correspondence, because similarity of function often results from
adaptation to common environmental demands, not common ancestry. As
a result, the idea of "functional homology" has long been thought suspect,
though there are ways to recover a coherent notion of homology of function
(Love 2007a). However, this strategy assumes that a functional trait can be
defined in terms of its activity (what it is) rather than use (what it is for). Is the
notion of a "conserved mechanism" inherently problematic because of its
reliance on functional (what it is for) individuation (Love 2018)?

The short answer is no, but the route to that answer requires that we examine
reasoning practices related to how biologists think about mechanisms and
make attributions of conservation. There is a critical link between our understanding of what a conserved mechanism is and the criteria for determining
whether a mechanism is genuinely conserved. Standard criteria for identifying
homologues include: (1) special quality, (2) similarity of structural detail, (3)
relative position in the body with respect to other traits, and (4) embryological
origin (Remane 1952). Each of these is neither necessary nor sufficient since
they can manifest due to other factors or fail to do so despite homology
obtaining; many homologues do not share the same embryological origin
due to developmental system drift (DiFrisco 2023; Haag 2014). Thus, the
criteria are not definitional but pertain to kinds of evidence that can support
a homology claim.

What might count as criteria for a conserved mechanism? We can use our
ecumenical characterization and ask whether its features can be mapped onto
the standard criteria for homology.

(1) Some constituents of a mechanism correspond to the criterion of special
quality. If the same key constituents are identified as standard components (e.g.,
Wnt proteins), then this is a particular feature (e.g., an initiating signaling
molecule) that is pertinent to determining whether something is conserved
(i.e., the Wnt signaling pathway). This helps account for why many conserved
mechanisms are named in terms of constituent molecules with these special
qualities (e.g., Wnt, Hedgehog, *Hox*, etc.).

(2) Organization of a mechanism corresponds to similarity of structural
detail. If a mechanism is organized in similar ways, in terms of which
families of molecules interact with one another (e.g., in the segment polarity
network), then this is another potential indicator of conservation. It supple-
ments the first criterion because many of the molecules whose name labels
a conserved mechanism are involved in other molecular processes that do
not exhibit the same organization among participating constituents.

(3) Spatiotemporal context of a mechanism can be mapped onto the criterion of relative position. The conserved mechanism of establishing A-P axes in the embryo using *Hox* gene expression occurs at distinct temporal junctures and spatial regions during development (e.g., early in the entire embryo, later only in appendages).

How should we deal with the fourth element that is pivotal in generating our conceptual puzzle ("to produce a specific behavior or phenomenon")? One plausible strategy is to recognize that claims about conserved mechanisms are not simple homology claims. That is, a judgment of correspondence for mechanisms across diverse taxa does not hold under every variety of form and function. Instead, our mapping between the traditional criteria for homology and those for conserved genetic mechanisms reveals a distinctive concept:

> *Conserved mechanisms are shared, derived traits composed of particular constituents, organized in a specific way, and found in delimitable spatiotemporal contexts where they manifest a stereotypical behavior or phenomenon.*

This is different from the standard understanding of homology because there are substantive claims about the sameness of constituents, organization, context, *and* function involved. Importantly, this distinct characterization underwrites the role of conserved mechanisms in securing explanatory generality from investigating model organisms. A typical judgment of homology would not necessarily license this kind of inference, such as conserved mechanisms for arthropod segmentation and vertebrate somitogenesis (Damen 2007).

Our characterization retains several advantages of standard homology judgments. First, we can still talk about sameness amid evolutionary modifications across taxa, accommodating variation in a mechanism's composition (form) or what it is for (function). Although some types of constituents are necessary for a conserved mechanism, there can be variation in the number of components and intensity of activities for each type. For example, the *number* of Wnt proteins involved in segment formation is variable across protostomes even though the *interactions* between Wnt and Hedgehog pathways is maintained (Janssen et al. 2010). Second, the requirement of specific organization need not imply identical organization across taxa. One example is the diverse rearrangement of gene interactions in conserved mechanisms related to the initial patterning of insect embryos (Chipman 2015). Arthropods show strong mechanistic conservation in the segment polarity network even though there are substantial patterns of divergence in upstream mechanisms operating earlier in development (Damen 2007). Third, spatiotemporal contexts and functional individuation can be treated at different levels of abstraction to establish correspondence. A reproductive signaling pathway can be considered conserved even though it produces outcomes of larval formation in soil worms, metamorphosis in insects, and puberty in mammals (Antebi 2013). At a higher level of abstraction, larval formation, metamorphosis, and puberty are all outcomes related to *reproduction*. Similar reasoning permits identifying conserved mechanisms across spatiotemporal contexts, such as the coordinated expression of *Hox* genes in establishing major body axes early in development and individual appendage axes later in development.

3.3 Payoff and the Problem of Deep Homology

We are now positioned to appreciate the payoff of our solution. Evo-devo research questions are addressed by an intensive experimental examination of conserved developmental mechanisms ("sameness") in model systems, which are compared in a phylogenetic framework to isolate *differences* in those mechanisms *relevant to* the origin and evolution of specific traits (Section 2.3). For example, what is the evolutionary history of segmentation in metazoans? Given that the three superphyla of Bilateria (Deuterostomia, Ecdysozoa, and Lophotrochozoa; see Figure 2) include both segmented and unsegmented taxa, how many times has segmentation evolved (Davis and Patel 1999; Minelli and Fusco 2004)? Comparing conserved and divergent features of segmentation mechanisms across different superphyla is crucial to answering this question. Studies of segmentation in leeches can be strategically informative (Kuo and Lai 2019; Weisblat and Kuo 2014). Unlike the other two superphyla that include standard model organisms, our knowledge about developmental mechanisms within Lophotrochozoa is more limited, which provides a major motivation to study leeches to secure comparative information about segmentation mechanisms (Weisblat and Kuo 2014).

These types of comparisons can be made in a more fine-grained manner in clades where more evo-devo model systems have been studied. The segment polarity network (Section 3.1) is only one complex mechanism in a suite of stages of gene expression that establishes a segmented animal. In *Drosophila*, molecular gradients are established by maternally expressed genes. Then zygotically expressed "gap" genes subdivide the embryo into four broad regions, followed by "pair-rule" gene expression that "stripes" the embryo into seven transverse parasegments along the A-P axis. Our familiar segment polarity network enters the story at this point, after which *Hox* gene expression confers identities on different segments (Hughes and Kaufman 2002). When these different stages of gene expression are compared across arthropods (e.g., crustaceans, insects, myriapods, and chelicerates), biologists have identified a high degree of variability in the mechanisms involving maternally expressed genes, gap genes, and pair-rule genes. However, the segment polarity network and its resulting parasegment units are highly conserved (Damen 2007). This makes it possible to construct plausible scenarios for how these mechanisms have evolved over time, exemplifying a "lineage explanation" (Calcott 2009). These evolutionary patterns in conservation can be teased apart further. Although the segment polarity network is highly conserved across arthropods throughout most body regions, the anterior-most segments in what will become the head exhibit such radical changes that the network's identity breaks down because of alterations to the constituents and organization of the mechanism (Lev and Chipman 2021).

There is another form of payoff from our analysis of conserved mechanisms that involves developing the idea further. A narrower notion – character identity mechanisms (ChIMs) – emerges from specifying that the phenomenon produced by the mechanism is a character (DiFrisco et al. 2020). This elaborates prior theorizing about how conserved identity networks underlie homologous characters (Davidson and Erwin 2006; Wagner 2007; Wagner 2014). The basic idea of an "identity mechanism" is that character identity – the status of an appendage as a wing rather than a leg or antenna – is not just a morphological fact but also something controlled and explained by distinctive causal mechanisms operating in development.

ChIMs for features at different levels of organization (cell types, tissues, and organs) not only play this role in character individualization (i.e., conferring identity) but also exhibit a shared developmental architecture for fulfilling that role. ChIMs lie at the midpoint in causal "bowtie" patterns, where variable inputs converge on a modular identity mechanism that can switch on variable, divergent outputs (Figure 10). Signals that activate ChIMs are susceptible to evolutionary change while maintaining their initiating role. Further downstream, the effector mechanisms switched on by a ChIM are responsive to selection on the character state that matches a particular functional role (e.g., antennae). Diverging lineages are therefore expected to accumulate differences in these downstream genetic and developmental mechanisms, whereas the ChIM is more refractory to evolutionary change (Wagner 2014).

ChIMs possess a shared causal profile or constellation of features that tend to be co-instantiated: (1) active maintenance through positive and negative feedback among components, (2) a complex organization of multiple heterogeneous components, (3) the necessity of components not only to maintain the ChIM but also to yield the development of the associated character (demonstrable in knockout experiments), and (4) causal nonredundancy – they do not coexist with other causes that have the same effect. Together, these features imply that ChIMs are less replaceable than upstream signaling inputs and downstream

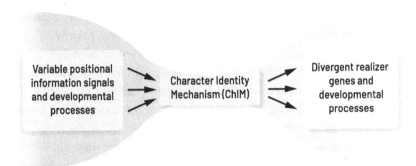

Figure 10 The ontogenetic location of character identity mechanisms. ChIMs are represented as the midpoint of a bowtie or hourglass pattern of development. Variable inputs converge on a conserved, modular mechanism that activates variable outputs. Redrawn.

Source: DiFrisco et al. (2023b).

effector mechanisms, being highly burdened or generatively entrenched (Novick 2019; Riedl 1978; Wimsatt 1986) and subject to strong stabilizing selection (DiFrisco et al. 2020). Importantly, they illuminate expectations about linkages between conserved mechanisms and their stereotypical outcomes.

Although there is more to say about the theoretical significance of ChIMs, one interesting consequence is that they suggest the commonly used notion of "deep homology" is problematic (DiFrisco et al. 2023b). This evo-devo concept emerged out of one of the most striking results of comparative developmental genetics: the same genes and pathways are *reused* repeatedly even in phylogenetically distant lineages (Held 2017). In exactly the way one would expect from evo-devo model systems, documenting similarities and differences in these pathways has helped dissect how novel traits originate through the reconfiguration and redeployment of existing genetic mechanisms (a theme we will return to in Section 4).

According to one account, deep homology describes, "the sharing of the genetic regulatory apparatus that is used to build morphologically and phylogenetically disparate animal features" (Shubin et al. 2009, 818). The first part of this characterization ("sharing of the genetic regulatory apparatus") can be understood as a conserved mechanism and the second part cashes out how it is individuated ("used to build morphologically and phylogenetically disparate features"). Thus, for two body parts to be deeply homologous, their developmental construction must depend on at least some of the same genetic regulatory features, and the body parts are not recognized as homologous or "would not be considered homologous by previous strict definitions" (Tschopp and Tabin 2017, 1). A classic example of deep homology is the involvement of homologous regulatory genes in the development of independently evolved eyes in *Drosophila*, mice, and humans (Quiring et al. 1994). The recognition of deep homologies leads to new questions: were these shared genes convergently recruited into different developmental processes, or were they parts of an ancient mechanism that patterned a more primitive character in the common ancestor (i.e., a homologue) (Shubin et al. 1997)?

Although the general fact of shared genetic components in phylogenetically distant lineages is consequential, the notion of deep homology has limited utility as a descriptive concept for comparative biology. The problem is a lack of precision. Two body parts can stand in a relation of deep homology while being homoplasies (i.e., nonhomologous structures), serial homologues (repeating structures in an individual, such as cervical vertebrae), or not even candidate homologues (e.g., beetle horns and insect legs). The term "disparate" in "used to build morphologically and phylogenetically disparate animal features" does not have a precise meaning in comparative morphological research, in contrast with "homology," "character identity," or "homoplasy" (Scotland 2010). Even if deep homology highlights how novel characters can arise from the modification of preexisting developmental resources, it does not shed light on what should count as a "novel" character or specify what constitutes those resources. With respect to the latter, deep homology is not distinguishable from ordinary gene sharing (where the same gene is used differently in different parts of the body – a metabolic enzyme in the liver versus a stable globular protein in the eyeball; Piatigorsky 2007) and pleiotropy (where the same gene has an influence on two or more distinct phenotypic traits). Are all characters that share genes or are influenced by a suite of genes deep homologies?[15] Additionally, the concept of deep homology does not have criteria for identifying a shared "gene regulatory apparatus" beyond gene homology (orthology); yet orthologous genes can have different causal roles in different mechanisms, especially at different times in development or different locations across regions in a developing embryo. Recall one of our clauses in the characterization of a conserved mechanism ("delimitable spatiotemporal contexts") and the segment polarity network breakdown for the most anterior arthropod segments.

These inherent difficulties with deep homology become visible from an explication of ChIMs, which are a refinement and elaboration of the notion of

---

[15] A similar concern can be raised for the notion of "partial homology." The continuity and overlapping genetic contributions to diverse traits across lineages seem to imply that no discrete determination of homology is possible. One can introduce further distinctions to existing concepts to produce increased contrastive resolution among the labels used to represent biological phenomena. See DiFrisco et al. (2023a) for an example of this approach focused on the concept of serial homology.

a conserved mechanism (DiFrisco et al. 2020). The ChIM model offers a notion of morphological character identity beyond similarity or "disparity," which clarifies what should count as novelty for body parts. It has a robust account of mechanism identity that is distinct from gene orthology, gene sharing, and pleiotropy, which clarifies what should count as novelty or sameness of identity mechanisms. Finally, the model provides a correspondence principle linking identity of characters to identity of mechanisms across different levels of organization, which facilitates moving from the recognition of novelty and the basis of character identity to a potential mechanistic explanation for the origin of new characters.

The precision afforded by ChIMs is crucial for ongoing research. First, ChIMs can be operationalized to facilitate empirical predictions and an experimentally decidable account of the identity of a biological character. They also make clear what it would mean for a new character identity to originate and how to test hypotheses about character origination, which is pertinent given that homologous characters can develop from different and diverse sets of genetic mechanisms (DiFrisco 2023). Second, many attempts to explain the evolutionary origins of novelty depend on appeals to the phenomenon of co-option (e.g., reusing a signaling pathway in a new context). Despite numerous patterns of co-option that are robustly supported by empirical evidence (Piatigorsky 2007; True and Carroll 2002), demonstrating that co-option has occurred is not the same as showing how co-option contributes to the explanation of new traits (Love and Wagner 2022). For the latter, we need a more delineated understanding of the level or unit of co-option (e.g., gene, gene network, or complex developmental process). Without this understanding, appeals to "deep homology" or related notions do not provide sufficient precision to discriminate among hypotheses based on different kinds of data, whether comparative transcriptomics or functional experimentation. Finally, assuming we have the requisite understanding of the relevant unit or level, there is a question of whether our experimental interventions can demonstrate how the co-option of specific mechanisms is sufficient to explain the novel character identity (Wagner 2001). Addressing this criterion requires detailed knowledge of developmental mechanisms in both the ancestral and derived conditions plus their instantiation in model systems where experiments are possible (recall the manipulation criterion for model organisms in Section 2). Otherwise, genetic perturbations cannot be unambiguously interpreted as evidence for how a new character arose evolutionarily.

We began this section with a puzzle about the very idea of a conserved mechanism. Through consideration of how we think about mechanisms generally and the criteria for detecting homology, we arrived at a characterization of a conserved mechanism that solved this puzzle and retained advantages relevant to evolutionary biological practice. Additionally, we observed how this solution yielded payoff for ongoing evo-devo research, such as for questions about the evolution of segmentation or the concept of deep homology. For the latter, this involved recognizing difficulties related to imprecision that affect the descriptive and explanatory use of the concept. This recognition emerged from the specific elaboration of conserved mechanisms found in the ChIM model, which is pertinent when the stereotypical phenomenon is a character, thereby constituting another part of the payoff: experimental traction that contributes to explanations of the origin of evolutionary novelties.

4 The Many Facets of Evolutionary Novelty

4.1 A Central Problem for Evo-Devo

In an oft-misquoted passage from *On the Origin of Species*, Darwin claimed that "How a nerve comes to be sensitive to light hardly concerns us" (Darwin 1964 [1859], 187). According to historians, the origin of new structures was not on his agenda: "Darwin explicitly disavowed theorizing ... about the origins of the most primitive eyes" (Lustig 2009, 113). It is not that he held it an unanswerable question ("several facts make me suspect that any sensitive nerve may be rendered sensitive to light"). Instead, he aimed to show how natural selection increased complexity. "Natural selection has converted the simple apparatus of an optic nerve merely coated with pigment and invested by transparent membrane, into an optical instrument as perfect as is possessed by any member of the great Articulate class" (188). However, many biologists have wanted to explain the origin of new structures, including features of eyes,[16] usually by attending to how novel variation in developmental genetic mechanisms originates in and through processes during embryogenesis. Morphologists and paleontologists were often at the forefront of these endeavors. They routinely attempted to account for the origins of higher taxa with reference to developmental transformations, such as the dramatic morphological effects of precocious sexual maturation in a larval form (Love 2007b).

---

[16] "Although these gradual–morphological progressions are logical and provide a powerful and visual way to imagine the stepwise evolution of complexity ... the developmental–genetic basis for how this morphological variation originates was not considered ... the origin of light sensitivity itself was not addressed, nor the origin of the cup structure or of the first lens material. ... using a gradual series of eyes as a model for how evolution proceeds is incomplete, because it assumes morphological variation without addressing the mechanisms leading to variation. How did light sensitivity originate? How did lenses or eye pigmentation originate? Answering these questions is critical for a complete picture of eye origins and evolution" (Oakley and Speiser 2015, 239).

On the landscape of contemporary biology, the problem of explaining the origin of novelties emerged as a key locus for evo-devo research and accounts for why evo-devo exhibits constellations of approaches from developmental genetics, morphology, and paleontology (Raff 2007). This problem represents a signature feature of evo-devo: "finding answers to what constitutes an evolutionary innovation ... and how developmental mechanisms have changed in order to produce these innovations are major issues" (Olsson and Hall 1999, 612). "Innovations are outside the scope of any current research program ... we see in the problem of innovation and the evolution of body plans a unique opportunity for [evo-devo] to develop its own independent identity as a research program" (Wagner et al. 2000, 822). These themes are echoed repeatedly in discussions of evolutionary novelty (Erwin 2021; Moczek 2008; Wagner and Lynch 2010). Apart from questions about the relationship of evo-devo to evolutionary theory (discussed in Section 1), we can characterize current investigations as tackling old evolutionary problems with new experimental and phylogenetic tools (Love and Raff 2003). This is especially evident when earlier discussions, like the origin of chordates (Garstang 1928), are set alongside recent evo-devo studies of the same topic (Lowe et al. 2015).

Despite a consensus that explaining the origin of novelty requires understanding development – "it is essential to include developmental mechanisms in the explanation of evolutionary innovations" (Wagner 2000, 97) – there are ongoing debates about what counts as a novelty and what causal factors best explain their origin. In addition to controversies surrounding the relative significance of gene regulatory network (GRN) changes in comparison to other factors (e.g., developmental plasticity), there remains a question about whether natural selection plays any explanatory role.

4.2 What is an Evolutionary Novelty?

Ernst Mayr defined an evolutionary novelty as "any newly acquired structure or property which permits the assumption of a new function," a definition that fits within the framework of the Modern Synthesis: "The problem of the emergence of evolutionary novelties then consists in having to explain how a sufficient number of small gene mutations can be accumulated until the new structure has become sufficiently large to have selective value" (Mayr 1960, 357). In contrast, a different definition has been in view for most work in evo-devo on the origin of novel structures. "A morphological novelty is a structure that is neither homologous to any structure in the ancestral species or [serially homologous] to any other structure in the same organism" (Müller and Wagner 1991, 243). Instead of functional properties ("selective value"), morphology in a phylogenetic context is foregrounded ("nonhomologous structures"); instead of mutations entering a gene pool, the focus is on the generation of qualitatively new variation. The growth of cladistic methods for phylogenetic reconstruction and emergence of molecular developmental genetics gave this structure-oriented definition operational traction (Section 1.1). And yet the definition relies on the concept of homology, which has exhibited its own share of semantic disagreements (Brigandt 2003; Gouvêa 2020; Novick 2018).

For some, these difficulties suggest that defining evolutionary novelty is a fruitless endeavor. Although the emphasis has been on novelty as a qualitative departure from an ancestral condition, the existence of a continuum between a qualitative difference and a quantitative variant makes it difficult to distinguish novelty from nonnovelty in absolute terms. Biologists draw the line on this continuum differently (Brigandt and Love 2012; Palmer 2012) and there are always precursors at lower levels (e.g., gene expression, cells, or tissues) for structures deemed qualitatively novel (Hall and Kerney 2012; Shubin et al. 2009). How should we interpret this diversity of definitions?

One possibility is that there is a theoretical lacuna. Perhaps a richer theory of what it means for something to be a character and have identity is required (DiFrisco et al. 2020; Section 3.3). Another interpretation involves challenging the assumption that we should have only one correct definition of evolutionary novelty. This challenge is based on shifting attention from defining the concept – delineating the criteria by which a term classifies entities ("categorization") – to characterizing the explanatory agenda associated with the concept (Brigandt and Love 2012). With this shift of attention, the meaning of the term "novelty" indicates explanatory expectations for studying the origin of morphological features, such as turtle shells (Figure 11; Kuratani et al. 2011; Lyson et al. 2013). For example, the criterion of nonhomology makes a trait's novelty a matter of what was historically present in a lineage. Forward-looking definitions characterize novelty in terms of developmental potential for future morphological variation and diversification (Wagner and Zhang 2011) or the capacity to transform an ecosystem's carrying capacity (Erwin 2012). The explanatory relevance of different causal factors varies. Thus, focusing on characterizing the explanatory agenda highlights questions about adequate explanations for novelties like the turtle shell, whether in terms of developmental mechanisms (e.g., the arrest of axial rib growth) or adaptive advantages in ecological contexts (e.g., burrowing and digging rather than protection).

On this perspective, there are many facets to the concept because of these different meanings. We can recover a broad characterization (rather than a narrow definition) of evolutionary novelties as derived body parts that

Figure 11 Different examples of the turtle shell, an evolutionary novelty.
Clockwise from top left: Red-bellied short-necked turtle (*Emydura
subglobosa*); Indian flapshell turtle (*Lissemys punctata*); Hawksbill sea turtle
(Eretmochelys imbricata); Galápagos tortoise (*Chelonoidis nigra*).

Adapted from: Wikimedia (Public domain). https://commons.wikimedia.org/wiki/File:
Turtle_diversity.jpg

usually lack homologous relations to structures in ancestral lineages and often
possess the potential for new functionality. However, the important feature of
this perspective is understanding explanatory expectations in different disciplinary contexts; we need to observe how the concept structures a problem
space (cf. Erwin 2021). This is often indicated by allied concepts used in
conjunction with a more operational definition of novelty. Discussions of
novelty as nonhomology emphasize the importance of distinguishing character identity (e.g., insect forewing) from character states (e.g., wing blade
versus protective cover) to clarify a sense of homology relevant to studying
the developmental genetic underpinnings for structural origination (Wagner
2014). Emphasizing the hierarchical level at which homology and novelty
apply plays a role in dissecting how mechanisms of gene regulation changed
evolutionarily to produce novel anatomical structures (Shubin et al. 2009).
Linking novelty to evolvability accents the relations between genotype and

phenotype (Pavlicev and Widder 2015).[17] Investigating the innovative ecological impact of a structural novelty makes natural selection germane to the explanation (Shirai et al. 2012). This strategy can be extended to include functional properties, such as behavior (Brown 2014a).

Additionally, typologies or classifications of kinds of novelty help to delineate explanatory expectations (Table 2). Müller (2010) offers a threefold typology of morphological novelty that emphasizes the epigenetic nature of developmental processes: type 1 – the primary anatomical architecture of a metazoan body plan; type 2 – discrete new elements added to an existing body plan; and type 3 – major changes to an existing body plan character. In alignment with this typology, Müller and colleagues offer a physical (epigenetic) explanation for type 1 and 2 novelties that is different from standard evo-devo approaches (Newman and Müller 2005; Newman et al. 2006). However, Wagner (2014) focuses on developmental genetic networks underlying character identity and advances a different, twofold typology. Type 1 corresponds to discrete new elements added to a body plan that require a new character identity network and type 2 corresponds to major changes in a character, such as a new character state.

Almudi and Pascual-Anaya (2019) describe a distinct fourfold classification because they hold that the other typologies conflate structures that are "truly new" with those that are progressive modifications of something in an ancestral lineage. Their typology excludes the latter and divides the former into four groups based on origination mechanisms: (1) combining existing structures to yield a new structure; (2) recruiting to a new location and modifying existing developmental genetic networks underlying character identity; (3) recruiting a cell type to new developmental contexts (in space or time) or the origin of new cell types; and (4) symbiogenesis (Almudí and Pascual-Anaya 2019). The diverse configurations of these typologies are driven by different expectations for what needs explanation and how best to explain it. Examples from one typology can be mapped into the space of other typologies, such as the origin of decidual stromal cells from endometrial cells as a new cell type for Almudi and Pascual-Anaya, which would be classified as a type 1 novelty for Wagner et al. (2019).

Can we say more about how the concept of novelty structures a problem space or agenda? One way to think about scientific progress is in terms of problem elaboration: "from problems to problems – to problems of ever

---

[17] The "genotype–phenotype map" can refer broadly to the diverse developmental trajectories from genetic resources and interactions to their phenotypic consequences or more narrowly to distinct theoretical or mathematical models of those developmental trajectories (Salazar-Ciudad et al. 2021).

Table 2 Different typologies of evolutionary novelty.

	Explanatory focus	Typology	Description	Example
Müller (2010)	Epigenetic interactions	Type 1	Primary anatomical architecture of a metazoan body plan	Chordate body plan
		Type 2	Discrete new elements added to an existing body plan	Appendage
		Type 3	Major changes to an existing body plan character	From fin to limb
Wagner (2014)	Developmental genetic networks	Type 1	Discrete new elements added to an existing body plan ("character identity")	Insect forewing
		Type 2	Major changes to an existing body plan character ("character state")	Wing blade or elytra
		Combination	Fusion of existing structures into a single unit	Insect wings (from tergal and pleural compartments)
Almudi and Pascual-Anaya (2019)	Generative mechanisms	Network recruitment	Modified GRN in new location	Beetle horns (from distal appendages)
		Cell type recruitment	New cell type or cell type in new developmental location	Decidual stromal cells (from endometrial stromal cells)
		Symbiogenesis	Fusion of different organisms	Bacteriomes in weevils

increasing depth" (Popper 2002 [1963], 301). That scientific problems can have "increasing depth" means they cannot be associated with a standard interrogative. Biological problems – such as how cells differentiate or how evolutionary novelties originate – are not single questions like "what genes make a difference in decidual stromal cell differentiation?" Biological problems constitute an agenda, a list of things that need to be addressed or multiple, interrelated questions that have been elaborated over time. Thus, one aspect of problem depth is the structure visible through extended historical debate (Hattiangadi 1978, 1979).

Another aspect of problem depth is epistemic heterogeneity (Laudan 1977), such as the difference between an empirical question (e.g., what is the distribution of phenotypic variation for a trait?) and a theoretical question (e.g., what is the relationship between genetic variation and phenotypic variation in evolutionary theory?). Depth also can be understood in terms of nested hierarchies, with problems containing subproblems or definable arrays of questions thought of as parts to the whole (Nickles 1981). These aspects of depth for research problems have been collected and systematized under the label "problem agendas" (Love 2008) and recent work offers a framework for unifying both intellectual and applied problems found in scientific inquiry (Elliott 2021).

What constitutes the epistemic history, heterogeneity, and hierarchy of a problem agenda like the origin of evolutionary novelty? Discussions in the recent history of evo-devo emphasize how novelties fall outside the standard models of evolutionary biology that concentrate on population genetic processes and quantitative genetic variation, largely ignoring the significance of development (Amundson 2005; Wagner et al. 2000). According to evo-devo practitioners, "it is essential to include developmental mechanisms in the explanation of evolutionary innovations [and] this is also the reason why developmental evolution makes an indispensable contribution to evolutionary biology" (Wagner 2000, 97). Whether this is true remains controversial (Hancock et al. 2021), but the historical controversy shapes the problem agenda through debate about the need for different disciplinary contributors to answer distinct and previously neglected or downplayed questions, including phylogeny and paleontology (to reconstruct character polarity and ancestral character states) and morphology (to determine the compositional identity of a feature and performance conditions for activities). Historical debate indicates some of the problem agenda structure, such as component questions and recognized criteria of explanatory adequacy.

The heterogeneity aspect of problem agendas – understood in terms of different types of questions – provides structure through expectations about the necessary intellectual contributions for answering them. Empirical questions ("what GRN controls appendage formation?") are answered differently from theoretical questions ("how is pleiotropy represented in a mathematical model?"); pattern questions ("what is the phylogenetic juncture for understanding jaw origins?") are answered differently from process questions ("how can changes in *cis*-regulatory binding sites contribute to heterotopy?"). Questions about cells differ from questions about anatomy; questions about the origin of features early in ontogeny (e.g., asynchronous cleavage of cells) differ from questions about those that occur later (e.g., metamorphosis). Different disciplinary approaches and methods are needed to address this heterogeneous set of questions in the problem agenda. That one discipline or approach focuses on some questions rather than others yields a division of labor and helps to organize different lines of investigation.

The final structural aspect of problems – hierarchy – highlights that the components of a problem agenda stand in systematic relations. This provides a template for how the various explanatory contributions can be coordinated and integrated; reflecting on the structure of the problem agenda reveals how an overall explanatory framework operates (Brigandt 2010; Love 2008). Some aspects of hierarchy can be cashed out in terms of abstraction and generalization. Questions that are more abstract ("how can complex traits be evolvable?") can be seen as higher in the problem structure hierarchy than others ("how can axial skeletal traits overcome developmental constraints due to pleiotropy?"). But since more concrete questions involve distinct biological processes ("how is GRN variation generated?" versus "how is epigenetic variation generated?"), the ability to offer an explanatory framework at the desired level of abstraction requires diverse methodological approaches. Although questions that are more general ("how do novelties originate in metazoans?") can be seen as higher in the hierarchy of problem structure than others ("how do novelties originate in mammals?"), more specific questions involve clade-level differences, which require that appropriately diverse taxa are studied, and the results judiciously compared (see Section 2.3). Otherwise, any explanatory framework will be overly fragmentary and less illuminating of how novelties originate.

Even when the investigative focus is on the origin of a particular novelty in one taxon, the necessity of coordinating diverse epistemic contributions remains. Because a morphological structure develops based on prior changes in gene transcription or cell migration, an account explaining the generation of this structure should be in terms of mechanistic interactions among lower- and higher-level features. Because an anatomical function (e.g., tetrapod limb movement) involves structures that articulate and interact in specific ways, an explanation of its origin is guided by relationships among that anatomical function's components. And, because the phylogenetic pattern preceding a novelty must be settled prior to assessing which developmental mechanisms contributed to the evolutionary transition, the architecture of the problem agenda not only requires different approaches (paleontology, phylogeny, and developmental biology) but also shows how contributions from these approaches should be integrated, thereby offering normative guidance for investigation. The hierarchical structure of a problem agenda provides a scaffold upon which to insert the relevant disciplinary contributions in an appropriate way.

4.3 Expectations for and Explanations of Evolutionary Novelty

Once the concept of evolutionary novelty is seen as representing an explanatory agenda, this facilitates making explicit the expectations or criteria of adequacy related to the problem space (Erwin 2021; Love 2008). These expectations indicate the necessary disciplinary contributions for an adequate explanation (see also Section 5). They include: (1) addressing both form and function, (2) being sufficiently general and abstract, and (3) exhibiting intricacy and balance to handle diverse questions and their interrelations.

(1) The emphasis on the adaptive modification of traits in neo-Darwinian population biology led to a neglect of questions about the origin of structure (Section 1.2). An emphasis on explaining the origin of morphology in the problem agenda of novelty corresponds to a correction of this functional bias. However, functional aspects surrounding the origin of new characters remain important (Brown 2014a), such as how phenotypic plasticity yields behavioral changes that affect morphology (Levis and Pfennig 2019; Moczek et al. 2011). Any adequate explanatory framework for novelty origins must address both morphology (form) and function. Since different disciplinary approaches engage form and function with a variety of theoretical and empirical methods, this criterion of adequacy prompts an integrated explanatory account of novelty.

(2) An adequate explanatory framework for the origin of new characters must be sufficiently abstract and general. The demand of abstraction derives from a need to integrate the necessary disciplinary contributions, such as the developmental generation of variation being investigated using methods from quantitative genetics, developmental genetics, epigenetics, and phenotypic plasticity. An aspect of the needed integration is an explicit articulation of the relations among levels of organization these methods concentrate on. The demand of generality requires that diverse characters in different clades are investigated using many methods, and that appropriate proxies for extinct taxa are utilized in experimental research with full knowledge of their epistemic limitations (Larsson and Wagner 2012; see Section 2.4). It also requires that

successful explanatory proposals for specific novelties (e.g., decidual stromal cells) be evaluated with respect to applicability to other characters.

(3) Adequate explanatory frameworks for novelty origins must exhibit sufficient intricacy and balance. Although this might seem counterintuitive (shouldn't we prioritize parsimony?), "intricacy" is about matching the heterogeneous questions in the problem agenda with corresponding answers. It goes together with the "balance" of an explanatory framework, which should handle empirical and theoretical questions, not neglect pattern questions for process questions, deal with lower levels of organization as well as higher levels, and address later moments in ontogeny in addition to earlier ones, as well as attend to both form and function.

These criteria of explanatory adequacy for the problem agenda show how the concept of evolutionary novelty goes beyond the categorization of traits into novel or nonnovel. The problem agenda is not preformed by the mere existence of the term but takes on shape and internal structure via deliberate and discursive explication (Love 2008). Agreement about the structure of the problem agenda is possible, even if it is not always settled which traits ought to count as novel, and thus accounts of the origin of a trait that fulfill the criteria of explanatory adequacy are genuine achievements, regardless of whether a trait is labeled a "novelty" by all definitions (Brigandt and Love 2012). By setting a problem agenda that structures ongoing inquiry, the concept of evolutionary novelty plays a fruitful epistemic role in evolutionary biology that is fully consistent with, if not supported by, the diversity of definitions for which characters count as evolutionary novelties.

What are some candidate explanations of evolutionary novelty? The most prominent strategy of research involves investigating how developmental genetic changes, especially the formation of new GRNs, contribute to novel morphological structures. "Evolutionary change in animal form cannot be explained except in terms of change in [GRN] architecture" (Davidson 2006, 29); "the evolution of development and form is due to changes within GRNs" (Carroll 2008, 30); "novelty requires the evolution of a new [GRN]" (Wagner and Lynch 2010, R50). These changes can involve duplications followed by the differentiation of paralogous genes (Gompel and Prud'homme 2009), modifications of regulatory interactions (Shirai et al. 2012), and the co-option of gene expression from one time or context to another (Piatigorsky 2007; True and Carroll 2002). For example, a shift in the spatial location of gene expression (heterotopy), rather than a shift in timing (heterochrony), was partially responsible for the origin of the vertebrate jaw (Shigetani et al. 2005). Some hold that heterotopy is the primary mechanism for the developmental genetic origination of novelties (Gompel et al. 2005). Thus, novel structures at higher levels of organization can arise from changes in conserved GRNs; they are explained by the recombination and redeployment of preexisting ancestral variation in developmental mechanisms, rather than because of novel genes. However, as we have seen (Section 3.3), in addition to isolating what modifications account for relevant gene expression differences yielding novelty, there are ambiguities in these claims that require closer attention to the nature of mechanisms that control trait formation (DiFrisco et al. 2023b).

Although there is agreement that evolutionary novelties arise from altered expression patterns due to changes in GRNs, some have argued that the evolution of protein–protein interactions leading to functional divergence in conserved transcription factors is also important (Lynch and Wagner 2008; Lynch et al. 2011). Other practitioners do not focus exclusively on developmental genetics and view different mechanisms as pertinent. For example, novel traits may begin as conditional structures that occur due to developmental plasticity (Moczek et al. 2011; Palmer 2012; West-Eberhard 2003). Regulatory modules are often reused and recombined in response to different environmental conditions, potentially generating novel phenotypic responses. Subsequent mutations in GRNs can permanently establish these structures via genetic accommodation or assimilation (Levis and Pfennig 2019).[18] Still other practitioners emphasize the significance of physical forces (e.g., fluid flow or differential adhesion) in generating morphological motifs (e.g., segmentation or tissue layering) under conditions of developmental plasticity early in metazoan evolution (Newman 2012; Newman et al. 2006). This explanation foregrounds epigenetic interactions at aggregate scales during development with novelties arising from combinations of physical patterning processes and cell properties under different environmental circumstances. Although this approach remains contested (Love and Lugar 2013), work on the genomic resources available at the origins of multicellularity is clarifying to what degree this hypothesis is feasible (Tweedt and Erwin 2015).

From an ecological vantage point, there are two distinct possibilities for the origination of a structural novelty: exaptation or developmental capacitance (Moczek 2007, 2008). Exaptation – a trait either originally selected for another purpose or a by-product of a different trait's formation that is exposed to a novel selective environment – presents the opportunity for a new adaptive function. Developmental capacitance represents processes that buffer against genetic or

---

[18] Genetic assimilation describes the process whereby a trait that was originally triggered by the environment loses its environmental sensitivity and becomes expressed constitutively in a population. Genetic accommodation describes the process whereby selection on environmentally triggered traits modifies heritable variation related to a trait, leading to increased, decreased, or different forms of plasticity.

environmental variation until a threshold for restraint is reached and breached. The potential was always there but remained cryptic until exposed (e.g., Reed et al. 2007). For both possibilities, the emphasis is on what conditions make it possible for natural selection to act and thereby consolidate the new trait via its functionality.

4.4 Evolvability: Conceptual Roles and Dispositions

Discussions of evolutionary novelty are closely allied to the study of evolvability – the capacity of a population to produce and maintain evolutionarily relevant variation. These studies concentrate on variational properties of genotypes, phenotypes, and the relationship between them to better understand evolutionary potential on different timescales (Hansen et al. 2023; Love et al. 2022). Researchers aim to disentangle whether the causes of evolutionary patterns arise from variational properties of traits or lineages rather than selection and ecological success. Achieving this would assist in debates surrounding the adequacy of evolutionary theory and calls for its augmentation (Section 1). Therefore, a brief consideration of the concept is worthwhile.

First, like evolutionary novelty, evolvability is conceptualized differently across fields of inquiry (Nuño de la Rosa 2017). Some researchers attribute evolvability to populations and understand it as the ability to respond to selection (Houle 1992), whereas others attribute it to organisms and understand it as the capacity to generate phenotypic variation (Kirschner and Gerhart 1998). One philosophical response to this situation is to identify a core meaning for evolvability. Differences in conceptualization are then understood as variations on this primary meaning, such as "the joint causal influence of ... internal features [of populations] on the outcomes of evolution" (Brown 2014b, 549). However, it is difficult to specify what counts as an internal feature of a population (Love 2003).

A different response is to analyze what these different conceptualizations accomplish in scientific reasoning (Villegas et al. 2023). This response assumes that the variation in conceptualization plays a functional role. Complementary possibilities for these roles include: tracking methodological approaches to a phenomenon of interest, representing distinct scientific aims (within or across disciplines), and locating commitments about a concept within a set of theoretical assumptions or with respect to its range of application (e.g., is it intended to apply only under certain circumstances?). An examination of scientific activities (e.g., setting a research agenda, characterization, explanation, prediction, and control) where the concept of evolvability plays a role in evolutionary inquiry helps to illuminate investigative and explanatory practices. For example, predictive accuracy predominates in evolvability studies from quantitative genetics (Hansen and Houle 2008), but the concept sets a research agenda in evo-devo, especially as something that requires characterization and explanation (Brigandt 2015a; Hendrikse et al. 2007). Additionally, resources for bridging different approaches can potentially help synthesize findings about evolvability across disciplinary boundaries (Villegas et al. 2023).

A second issue arises from evolvability being a disposition. A dispositional property is a capacity, ability, or potential to exhibit some outcome, which one can possess without displaying it. They are common in biology (Hüttemann and Kaiser 2018) and especially abundant in evo-devo, such as the capacity to exhibit distinct traits under different environmental circumstances (phenotypic plasticity; see Austin 2017).[19] Evolvability, as a disposition, can but need not manifest in a higher rate of evolution because two populations with an identical ability to evolve may exhibit different evolutionary outcomes due to chance or because the two populations are exposed to environmental conditions with different regimes of natural selection.

The dispositional nature of evolvability matters; philosophical distinctions about dispositions have scientific implications (Brigandt et al. 2023). By wielding basic distinctions about dispositional properties (e.g., background conditions, bearer, causal basis, intrinsic versus extrinsic, or probabilistic versus deterministic), one can address questions about what kind of disposition evolvability is in these different contexts, and how different conceptions and disciplinary approaches are related. What is the difference between an individual entity and a population of entities bearing the disposition? What is being measured when studying evolvability empirically (causal basis, disposition, manifestation, or something else)? Are contributors to the causal basis of evolvability only intrinsic to the bearer of evolvability?

Taking up this last question, a focus on development as the causal basis of evolvability tends to accent features "internal" to organisms: "The evolvability of an organism is its intrinsic capacity for evolutionary change. ... It is a function of the range of phenotypic variation the genetic and developmental architecture of the organism can generate" (Yang 2001, 59). But is evolvability always an intrinsic capacity? Recognizing that the internal/external distinction is a matter of theoretical perspective, many scenarios suggest evolvability can depend on external factors. If a phenotypic trait is the bearer of evolvability,

---

[19] Some have claimed that standard evolutionary theory (understood primarily as population genetics) only utilized categorical properties and evo-devo's concentration on dispositional properties constitutes a necessary addition to the ontological commitments that underlie evolutionary explanations (e.g., Austin 2017). This connects with discussions in Section 1.2 and goes beyond it to concerns from the metaphysics of science that I ignore here.

then extrinsic features may compose part of its causal basis to evolve, such as pleiotropic relations to other traits or the frequency of the trait in the population. If the bearer is a population or taxon, then interactions with abiotic or biotic entities in the environment can have an impact on evolvability (Love 2003; Sterelny 2007); a taxon's extinction rate can depend on its geographic range defined by landscape topography or ecological diversity (Jablonski 1987). An overarching lesson from these considerations of why or when evolvability might have an extrinsic, relational causal basis is that these judgments depend on the specific conception of evolvability being used and the role it plays in different scientific activities (Brigandt et al. 2023; Villegas et al. 2023).

4.5 Quo Vadis?

Where does this leave us? What next steps might researchers take given the many facets of evolutionary novelty, including their relevance to evolvability? That explanations of novelty require multiple disciplinary contributions will be explored further in the next section. I close this section with three dimensions – conceptual, empirical, and theoretical – that hold promise for future inquiry into the origins of evolutionary novelty.

Inquiry related to the conceptual dimension can advance our understanding by intentionally exploring different definitions of novelty. Many researchers operate with a single preferred conception of novel structures. If we think of these conceptions as different models (Wagner 2014), then switching between them focuses our attention on different properties of biological systems and encourages analyzing the evolutionary significance of different causal factors. The key is to recognize that distinct criteria of explanatory adequacy accompany these different models and foreground some factors while relegating others to the background. Explanatory accounts derived from different definitions may be complementary rather than competing, especially those that appeal to different kinds of developmental mechanisms or adaptive benefit. Progress with respect to explaining the origin of novel structures is nurtured when explanatory aims are articulated precisely, their scientific significance is widely comprehended, and the standards that govern the structure of an adequate and integrated explanatory framework are as explicit as possible (Brigandt and Love 2012; Erwin 2021).

Inquiry related to the empirical dimension can advance our understanding of the origin of novelty by increasing the manipulative capacity of experimentation and augmenting our inferential capacities. The ability to precisely change genetic variables and ascertain their phenotypic effects in evo-devo model systems has long played a role in explaining novelty (Section 2.3), but new technologies, such as CRISPR-Cas genome editing and RNA-seq to characterize gene expression, provide more discriminating tests of hypotheses for putative novelties in evo-devo models. Yet, as noted, explaining the origin of novel structures requires a combination of disciplinary approaches and therefore empirical advances are needed on multiple fronts. Morphological and paleontological investigations have illustrated this dramatically for the tetrapod limb (Shubin et al. 2006). Integrating developmental genetic and physico-chemical mechanisms can augment our understanding of how new traits originate (e.g., Sheth et al. 2012). Finally, increasing phylogenetic resolution can change how existing empirical data from various approaches bear on explanations of novelty. Investigating novelties that represent the best available juxtaposition of these empirical tools will generate the deepest insights, which may mean choosing unexpected model systems (e.g., Wagner et al. 2014).

Inquiry related to the theoretical dimension can advance our understanding by probing more quantitative dimensions of the genotype–phenotype map and drawing out abstract generalizations across disparate systems (Hansen et al. 2023). The former provides a bridge to the rich tradition of population genetic theorizing that has sometimes been perceived as antagonistic to evo-devo's developmental orientation (Nunes et al. 2013). It also introduces different concepts into explanatory accounts (e.g., pleiotropy or epistasis) that augment our understanding of the origin of qualitatively distinct variation at particular phylogenetic junctures (Pavlicev and Wagner 2012; Rice 2012). More abstract generalizations can be derived from forms of theorizing by identifying commonalities across taxa and levels of organization, such as shared network architecture in metabolism and gene regulation (Wagner 2011).

Furthermore, this theorizing need not be concerned only with population-level modeling. The increased precision of ChIMs over deep homology (Section 3.3) can help resolve issues of character identity and novelty (DiFrisco et al. 2023b). It makes the question of whether a case of gene sharing is homology or co-option experimentally decidable rather than a comparison of transcriptomic or morphological similarity. Similarity of gene expression is too weak a criterion to distinguish between serial homology and a novel character arising from the co-option of several genes or other developmental mechanisms (DiFrisco et al. 2023a). Moreover, an explicit ChIM model can guide the interpretation of experimental results beyond standard appeals to deep homology, gradually raising the standards of evidence that facilitate discriminating among different hypotheses. We thereby secure a richer and more empirically precise notion of co-option, overcoming a key barrier to experimental demonstrations that rely on standard model organisms, and establish what is needed to carry out synthetic experiments that help to causally explain evolutionary novelty. Overall, combinations of advances in conceptual, empirical, and theoretical dimensions harbor tremendous promise for providing deeper and more adequate explanations of the origin of evolutionary novelties.

5 Interdisciplinarity and Explanation in Evo-Devo

5.1 Interdisciplinarity

In Section 1, we saw that evo-devo can be characterized in several ways. A commonality across these characterizations is that researchers from different disciplinary backgrounds, who use an assortment of methods and approaches, see themselves as working within evo-devo (sometimes to the exclusion of one another). Computational modeling, developmental genetics, experimental embryology, morphology, paleontology, and systematics are a small sample of these disciplinary backgrounds. They share the aim of explaining the origin and evolution of form or structure, which fosters a common commitment to incorporating an understanding of developmental mechanisms that generate phenotypic variation into evolutionary theorizing. The criteria of adequacy for the problem agenda of evolutionary novelty demand multidisciplinary explanations (Section 4).

Is there a difference between multidisciplinary and interdisciplinary? Etymologically, interdisciplinarity means "between disciplines," where disciplines are standardized fields of scientific research (Repko 2008). The intuition is that something distinctive and unavailable to standard areas of study emerges from interdisciplinarity, such as a more adequate or general explanation. Some have argued that "multidisciplinary" refers to research that brings disciplines together for a particular purpose while retaining their distinctness, whereas "interdisciplinary" more permanently integrates multiple disciplines to produce a new discipline (Collins 2002). The latter is one way that evo-devo has been characterized (Raff 2000), and the growth of institutional infrastructure, such as professional societies, new research journals, dedicated funding, and textbooks capture a sense in which evo-devo is interdisciplinary (Moczek et al. 2015). However, other strands of research suggest multidisciplinarity, with disciplines organized transiently around the solution of specific research questions, such as the origin of novelties in the fin–limb transition (Brigandt 2010). The institutional structure of contemporary science might be unaffected even though morphologists from a Department of Ecology and Evolution, paleontologists from a Department of Earth Science, and developmental biologists from a Department of Genetics, Cell, and Development collaborate to address research questions about the evolution of development.

Fortunately, there is no need to adjudicate between these conceptions. Regardless of the depiction of evo-devo, all involve multiple participating fields of study with multidisciplinary and interdisciplinary elements present. This aligns with inquiry into interdisciplinarity, which has identified the presence and articulation of one or more complex problems as a prerequisite (Repko 2008). Consensus has coalesced around a characterization of interdisciplinary research: "A mode of research by teams or individuals that integrates information, data, techniques, tools, perspective, concepts, and/or theories form two or more disciplines or bodies of specialized knowledge to advance fundamental understanding or to solve problems whose solutions are beyond the scope of a single discipline or area of research practice" (National Academy of Sciences 2005, 39).

The evolution of development and the developmental basis of evolution fit within this ambit.

5.2 Complex Phenomena

That evo-devo is interdisciplinary is beyond dispute (Love 2021). Why it should be interdisciplinary is another question. One answer is that the complex problems that characterize evo-devo as a community of inquiry derive from the complexity of evolutionary and developmental phenomena under scrutiny, both with respect to spatial levels of organization and temporal scales on which causal interactions across these levels occur. Consider again the shared evo-devo commitment: "In order to achieve a modification in adult form, evolution must modify the embryological processes responsible for that form. Therefore an understanding of evolution requires an understanding of development" (Amundson 2005, 176). What is required to "understand development"? When researchers emphasize the need to understand mechanistically how phenotypic variation is produced, one aspect of this is comprehending the dynamics of development at multiple levels of organization (Brooks et al. 2021). Embryogenesis begins in many metazoans with a single cell from which multicellular aggregates (tissues) are formed; these tissues, in turn, compose a functioning organ or anatomical feature. Differential gene expression leads to changes in cell fate that facilitate different kinds of tissues, organs, and anatomy. Sometimes the movement of tissues or mechanical pressure from anatomy induces a change in gene expression. An understanding of development for evolution requires dissecting the variety of causal mechanisms operating during ontogeny (Baedke 2021; Brigandt 2015b).

This mechanistic variety yields hierarchies of developmental structure and process that can be described as complex in two ways: compositional and organizational (Love 2006). Compositional complexity refers to the material constitution of characters (part–whole relationships), both in terms of the number and types of components. Larger numbers of cells and a higher number of cell types that comprise a character can be understood as increases in complexity. Organizational complexity refers to the causal relations that obtain between components. One aspect is the degree of aggregativity – how much causes act in a linear fashion – with increasing non-aggregativity reflecting increased complexity (Wimsatt 1997). Another aspect is the number and kind of structural arrangements exhibited by components and component types that are relevant to yielding causal outcomes. Higher numbers and kinds of structural arrangements count as more complex.

Thus far, we have considered compositional complexity and organizational complexity in developmental time within a single generation. However, investigating the evolution of development and developmental basis of evolution also involves considering evolutionary time across generations (Brown 2021; Calcott 2009). This expansion of temporal scales indicates unambiguously that complexity is a multifaceted attribute of the phenomena of interest studied by evo-devo researchers. To understand the origin of neural crest cell migration (the evolution of development), GRNs (organizational complexity) involved in the folding of the neural tube (compositional complexity) need to be detailed both in an ancestral and derived lineage (two distinct developmental times), so that we can identify how changes in gene expression in the lineage (evolutionary time) yielded the capacity for the detachment and migration of neural crest cells. Aspects of compositional complexity, such as how neural crest cells compose different tissues, are subject to transformations through evolutionary time (Table 3). To understand how the genotype–phenotype map facilitates the evolvability of a trait (the developmental basis of evolution), properties such as pleiotropy (organizational complexity) and how they relate to distinct developmental modules (compositional complexity) need to be theoretically articulated in ancestral and derived taxon representatives (different developmental times), so that we can identify what happens across generations (evolutionary time) through selection experiments, simulation modeling, or fossil record data (Love et al. 2022). Organizationally complex properties such as pleiotropy are subject to evolutionary transformation in a lineage.

One further element of complexity for evo-devo phenomena is relevant for interdisciplinarity: taxonomic scope. The properties and processes at different spatial and temporal scales are instantiated in a diversity of taxa across the tree of life. Compositional and causal relationships differ across taxa, such as numbers of cell types or variation related to generation time. Given that some disciplinary structure in biology is taxon-specific (e.g., entomology or herpetology), interdisciplinarity in evo-devo involves more than coordinating approaches from cell biology, development, and paleontology to address this complexity. The aim of formulating

Table 3 Possibilities for complex developmental and evolutionary phenomena. Hierarchies can be either compositional (part–whole) or procedural (control or process dependence), concern either form or function features, and occur in developmental time (within a single generation) or evolutionary time (across generations). Example 1: A compositional form hierarchy in developmental time for organ origination could be cells aggregating into tissues, thereby allowing tissues to aggregate into organs: specific form features are nested within ("subparts of") the morphological novelty during ontogeny within a single generation. Example 2: A procedural function hierarchy in evolutionary time for the origin of neural crest cell migration could be gene expression involved in the folding of the neural tube originating prior to gene expression involved in the detachment or migration of neural crest cells: specific function features must activate serially or jointly prior to the operation of the organismal innovation during the evolutionary process across generations. Source: Love 2006.

Hierarchy	*Form*	*Function*	Time
Compositional	Example 1		*Developmental*
			Evolutionary
Procedural			*Developmental*
		Example 2	*Evolutionary*

explanatory accounts of the evolution of development or developmental basis of evolution that hold generally across regions of the tree of life requires recognizing the contribution of taxonomic scope to developmental and evolutionary phenomena.

Biologists acknowledge that this complex reality – wherever and however it is distributed taxonomically – undergirds the rationale for interdisciplinarity: "Because the mechanisms of each trait of interest are manifested at lower levels of biological organization and the significance of a trait is only apparent at higher levels, understanding a given trait usually requires the simultaneous use of molecular, cellular, organismal, population and ecological approaches" (Feder and Mitchell-Olds 2003, 649). The multifaceted complexity of development that needs to be comprehended if one is to incorporate how phenotypic variation is generated and what possibilities for variation are available (i.e., variability) requires interdisciplinarity in evo-devo. The different levels of organization and temporal scales on which interactions occur in different taxa are the province of many life science disciplines (as well as physics, chemistry, and engineering). Adequate explanations of different dimensions of the evolution of development or developmental basis of evolution will not emerge from a single disciplinary approach. However, that the phenomena of interest in evo-devo research exhibit a multifaceted complexity and demand interdisciplinarity does not yet illuminate *how* researchers coordinate their diverse methods and approaches across disciplines or achieve a synthesis or integration of their results, especially given that evaluative standards for what counts as data or a good explanation can vary across fields of study.

5.3 Structured Problems, Interdisciplinary Integration

One of the most philosophically conspicuous features of interdisciplinary investigation is that it does not arise in response to the aims of theory construction and confirmation. The request originates from complex problem domains that elude scientific explanations derived from disciplinary approaches, as would be expected for complex phenomena. "A common argument favoring interdisciplinary research refers to the nature of problems that science is supposed to help solve ... their solutions require the combined effort of many traditional disciplines. ... Scientists need to integrate all these facts to solve the problem" (Hansson 1999, 339). These complex problem domains pick out suites of research questions that can be described as *problem agendas* (Section 4.2). Although these questions are addressed in part through the application of existing, well-confirmed theories, it is the complicated interrelations among questions within a problem agenda that are responsible for spurring the call to interdisciplinary research. Thus, the complex phenomena of development and evolution are partitioned into problem agendas (e.g., the nature of evolvability or the origin of novelties) that themselves are complex features of scientific reasoning.

A pressing concern is whether problem agendas have features that facilitate the coordination of disciplinary approaches (where a contribution is made) and their evaluation (the standards for assessing contributions). Coordination can be understood in terms of the structural depth of a complex problem domain. The interrelated suites of research questions that compose a problem agenda exhibit organizational architecture in terms of different kinds of research questions ("heterogeneity") and hierarchical relationships among questions – questions contain sub-questions, or one question depends on answers to other questions. These features were illustrated earlier for the problem agenda of evolutionary novelty (Section 4.2).

Heterogeneity and hierarchy operate concurrently to yield structure that can both coordinate disciplinary approaches, including how they should be integrated, and contribute to their evaluation. The latter is especially important since there is not agreement about solutions to general questions, such as how novelties originate: "epigenetic mechanisms, rather than genetic changes, are the major sources of morphological novelty in evolution" (Newman et al. 2006, 290); "evolutionary change in animal form cannot be explained except in terms of change in gene regulatory network architecture" (Davidson and Erwin 2006, 29). The hierarchical structure of research questions suggests configurations of inquiry. More concrete questions involve distinct biological processes, which are the province of different fields of study (comparative developmental genetics versus bioengineering) and must be weighed for their relative significance in achieving an adequate explanatory framework at the desired level of abstraction. More specific questions involve clade-level differences, which require the study of diverse taxa and the assiduous comparison of results.

The focus of one discipline or configuration of several fields on some questions rather than others creates a fruitful division of labor and organizes different lines of investigation in terms of the kinds of questions they concentrate on. For example, because morphological structure develops based on prior changes in lower-level traits, an account explaining the generation of morphology must capture these mechanistic interactions. And because the phylogenetic juncture for character origins must be settled prior to assessing which developmental mechanisms contributed to the evolutionary transition, the architecture of the problem agenda not only requires different approaches (paleontology, phylogeny, and developmental biology) but also points to how contributions from those approaches should be integrated, such as through answering particular kinds of questions. Thus, both heterogeneity and hierarchical structure in a problem agenda offer a template that normatively guides relevant disciplinary contributions to their appropriate points of insertion.

Beyond the coordination of diverse epistemic contributors, there remains a question about the standards for assessing contributions. These can be understood in terms of the agenda's associated criteria of explanatory adequacy, which help to organize research because ongoing inquiry is directed at fulfilling these criteria. In Section 4.3, we observed three criteria for explaining the origin of new characters: addressing form and function, sufficiently abstract and general, and exhibiting intricacy and balance. Importantly, what counts as sufficiently abstract and general or intricate and balanced for an explanatory account is subject to scientific change through history (Love 2015a). Dimensions of problem agendas, such as heterogeneity and hierarchy, take on different shapes and contours as research uncovers novel empirical results or develops new theoretical perspectives. Although there is continuity, such that researchers can recognize how past investigation is linked to current inquiry, these changes can be substantial and include whether one problem agenda is prioritized over others.

Sometimes changes are related to developments within disciplinary approaches themselves. The availability of new methods has been crucial to modern configurations that we observe working collaboratively in evo-devo.

For example, molecular genetic tools that were forged in the context of model organisms to better understand development (e.g., *in situ* hybridization to detect spatially localized gene expression) became central to interrogating the nature of homology (Holland et al. 1996). Theoretical advances in phylogenetic reconstruction occurred within systematics and facilitated the use of molecular data, which clarified character polarity and led to better hypothesis testing about the evolution of developmental features (Telford and Budd 2003). Additionally, changes within one discipline can have an impact on others. Advances in the application of an engineering perspective to embryos grew out of adopting standards from molecular developmental genetics for establishing causes in developmental processes through the manipulation of experimental variables. This alignment of standards across disciplinary approaches then implies that explanations of the origin of novelties must integrate both genetic and physical causes (Love et al. 2017), thereby transforming the shape of the interdisciplinary problem agenda.

Thus far, we have concentrated on how complex natural phenomena prompt us to tailor our epistemology of scientific investigation to that reality. However, it is crucial to acknowledge that this directionality can be reversed. Features of scientific epistemology, such as methodological commitments or preferred explanatory approaches, guide how investigators study the evolution of development and developmental basis of evolution. Criteria of explanatory adequacy reflect epistemic values of biological researchers, such as generality, which hold independently of the complexity of evolutionary developmental phenomena. Molecular genetic methods used to identify gene expression relevant to GRN dynamics fostered a conception of developmental phenomena and their evolutionary significance that tended to ignore the potential contribution of protein–protein interactions (Lynch and Wagner 2008). Similarly, innovations in the quantification of morphology have made it possible to document and comprehend shape change in developmental sequences and through evolutionary time (Mitteröcker 2021). These epistemological and methodological aspects change how researchers study complex phenomena. Sociologically, different disciplines incentivize kinds of research, which can nurture some types of interdisciplinary collaboration and discourage others. In the 1980s, an intellectual environment interested in epigenetic dynamics fostered collaborations that included physical science (Oster et al. 1988). Subsequently, the centrality of GRNs downgraded attention to epigenetics: "Developmental complexity is the direct output of the spatially specific expression of particular gene sets and it is at this level that we can address causality in development" (Davidson and Peter 2015, 2). Only recently has combined attention to genetics and physics begun to resurface

(Love et al. 2017; Newman 2012). The complex phenomena have always been there, but the epistemological inclination to explore them has waxed and waned over time.

The tasks laid out by the different problem agendas found within the evolution of development and the developmental basis of evolution are daunting. However, several procedural lessons can be drawn about explanatory integration across disciplines. The relevant types of disciplinary contribution can occur in a piecemeal fashion, directed at specific research questions (as opportunity permits), rather than being cast at a global level ("the developmental basis of evolution"). The concreteness and specificity of these "local" research endeavors facilitate a more transparent picture of what intellectual contributions are needed for an adequate explanation; novelties at different levels of organization may require distinct explanatory ingredients in different combinations and thus intricacy and balance can be secured through a patchwork synthesis of local integrations of distinct disciplinary approaches. Successful interdisciplinarity coordination involves different integrative relations across fields and for delimited tasks or times (Brigandt 2010). Progress in evo-devo need not be measured by a consensus set of theoretical relations across all relevant fields of study. A good example of this piecemeal progress is found in a classic evo-devo problem: the fin–limb transition.

5.4 Interdisciplinarity and the Fin–Limb Transition

The transition from fins to limbs in the history of life is a long-standing evolutionary puzzle associated with the origin of tetrapods and vertebrate invasion of land (Figure 12) (Tanaka et al. 2021). Addressing the empirical and conceptual questions that compose this problem requires multiple disciplinary approaches, each with specialized concepts and methods: "the challenge is to continually synthesize knowledge gained from multiple perspectives into an ever more refined understanding" (Hall 2007, 151). The origin of the autopodium (hand/foot) – fins minus fin rays plus digits equal limbs – is informed by new fossil findings (matched with detailed morphological analysis) that shed light on the ancestral character state of the fin and facilitate more refined phylogenetic reconstruction (Stewart et al. 2020), the identification of shared features in the development of fin rays in fish and tetrapod digits (Nakamura et al. 2016), and functional morphological analyses of fish locomotion (Kawano and Blob 2013). Studies of the developmental variation and generation of digits is not confined to molecular genetics but includes approaches that specifically address physical dynamics (Onimaru et al. 2016). Different processes that involve

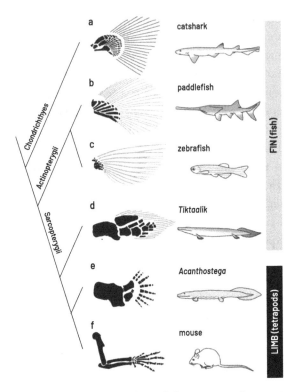

Figure 12 Phylogenetic representation of the comparative anatomy relevant
to the fin–limb transition. *Tiktaalik* and *Acanthostega* are extinct taxa.
Redrawn.

Source: Tanaka et al. (2021). Reproduced by permission of Springer Nature.

mechanisms across levels of organization include chondrogenesis (cartilage), osteogenesis (bone), apoptosis (programmed cell death), joint formation, postnatal growth, and regeneration.

There is compositional complexity in the form of fins and limbs, such as endochondral versus dermal bone (Nakamura et al. 2016), as well as organizational complexity in how this morphology is generated (endochondral bone arises from aggregations of mesenchymal cartilage cells later replaced by mineral bone; dermal bones mineralize directly from mesenchyme without a cartilaginous intermediary step). The skeletal arrangements in fins and limbs have distinctive patterns of size and shape that are observable in fossil and extant taxa. Compositional complexity extends to other tissues (e.g., blood vessels) and both cellular and molecular constituents (e.g., collagen). On a developmental timescale, organizational complexity results from common sources (e.g., lateral plate mesoderm) but distinct developmental mechanisms (e.g., differential patterns of *Hox* gene expression or alterations to epithelial morphogenesis). On an evolutionary timescale, two distinct developmental changes occurred in the origin of the tetrapod limb: a loss of fin rays and the expansion of the endochondral region of bone (tetrapod limbs contain no dermal elements). Additionally, there are interactions between compositional and organizational complexity, such as the evolution of developmental mechanisms relevant to fins and limbs.

Since the fin–limb transition is situated within the problem agenda of evolutionary novelty, we can observe a heterogeneity of question types, whether empirical ("what changes in gene regulation are responsible for spatial and temporal patterns of *Hox* gene expression?) or theoretical ("how is the functional demand of terrestrial weight-bearing managed in limbs with different skeletal arrangements?). Questions about the cellular level of organization differ from questions about the anatomical level; questions about the source of cell populations earlier in ontogeny differ from questions about morphogenetic trajectories of epithelial sheets later in development. Hierarchy in the sense of relations among component research questions can be detected in abstract questions ("how important are Turing-type reaction–diffusion models compared with GRN models for understanding digit origins?"), as well as more specific sub-questions ("how does the GRN of Bmp-Sox9-Wnt operate in fins and limbs?"). General questions ("how does endochondral bone expand in tetrapod limb evolution?") have counterparts that are less general ("how do mesopodial wrist elements emerge in the limbs of stem tetrapods?).

This abbreviated case study exemplifies how problem agenda structure helps to coordinate and integrate disciplinary contributions and their evaluation. A variety of investigative approaches from different fields of study are required and need integration within the structure of the problem agenda. The relative importance of differential gene expression and biomechanics needs to be weighed, as do morphological, paleontological, and systematic analyses. Progress derives both from empirical contributions (e.g., new fossils) and an enhanced integration of contributions, such as an increased interweaving of developmental genetics and physical dynamics for digits or the different trajectories of common developmental precursors in fins and limbs. Together, these integrated contributions are on track to yield a sufficiently abstract and general explanatory account that is reciprocally informative for other evolutionary novelties (e.g., the origin of vertebrate jaws), while maintaining an appropriate intricacy and balance that addresses heterogeneous questions with adequate empirical detail.

5.5 Final Reflections

Despite divergent perspectives on what evo-devo is and different formulations of the meaning of interdisciplinarity, there is no question that interdisciplinarity is central to its identity and practices. The primary rationale for why evo-devo investigations exhibit such a diverse mixture of disciplinary contributors is the complexity of the evolutionary and developmental phenomena under scrutiny. This holds for both spatial levels of organization – compositional complexity – and temporal scales on which causal interactions within and across these levels occur – organizational complexity. Biologists represent this multifaceted complexity in nature as structured problem agendas that organize diverse research questions and help to coordinate disciplinary approaches (where a contribution is made) and their evaluation (the standards for assessing the contributions). Both the complexity of the evolutionary developmental phenomena and this representational architecture can be observed in reciprocal interactions through history in the case study of the fin–limb transition.

Our analysis has both epistemological and sociological implications. Epistemologically, the complexity of phenomena and structure of problem agendas suggest that the enterprise of evo-devo is likely to yield fragmentary explanatory frameworks in response to different domains of problems. This implication is reinforced by the subtle feedback effects between problem agendas, preferred methodologies, and social structure in biology. It means that there should be caution in demanding a fully integrated and unified theoretical framework in evolutionary biology that incorporates both the evolution of development and the developmental basis of evolution. Although these frameworks will display different degrees of integration as answers to various questions within a problem agenda are identified and interconnected, it is unclear whether a single or small set of principles will be adequate to account for all the relevant phenomena. This does not mean that parallels across spatial levels or temporal scales should not be sought; instead, expectations should be tempered for the resulting knowledge that derives from interdisciplinary inquiry in evo-devo (Richardson 2022). Rather than a single, unified theoretical perspective, multiple models will typically persist as a feature of successful explanations offered by researchers.[20]

---

[20] Evo-devo also has epistemological implications for other fields, having spurred new forms of research on cognition (Ploeger and Galis 2021), culture (Charbonneau 2021), and language (Balari and Lorenzo 2021). Controversies surrounding these attempted reconfigurations of other research domains often reflect similar fault lines as observed in Section 1. Some have even argued that evo-devo should be generalized to all levels of phenotypic evolution (Salazar-Ciudad and Cano-Fernández 2023).

These epistemological implications also bear on the sociological structure we should expect in evo-devo. The coordination of disciplinary approaches and their evaluation is a difficult task, and one that is not conducive to the regular operation of scientific disciplines. Disciplinary structure in contemporary science serves several purposes that require a narrowing of methods and approaches to sufficiently constrain or "discipline" these units of scientific organization. Standards of evaluation in morphometrics differ from those in comparative developmental genetics. Maintaining loci where the full range of interdisciplinarity is on display is difficult, whether in professional meetings or peer-reviewed journals. Wherever more overt sociological disciplinary structure has emerged within evo-devo (e.g., the Pan-American Society for Evolutionary Developmental Biology in 2013–2014, or specific journals such *Evolution & Development*), there has been a constriction (intentional or not) of the kinds of research appearing in these outlets. Discipline formation involves a tightening of boundaries. Yet the kinds of work needed to adequately answer research questions within problem agendas related to the evolution of development and developmental basis of evolution occur in other disciplinary contexts and variegated publishing outlets. The same holds for funding evo-devo research, which comes from sources other than "The Evolution of Developmental Mechanisms Program" at the National Science Foundation (for example).

These epistemological and sociological implications of the interdisciplinarity of evo-devo remind us that the necessary integration among answers to research questions within problem agendas will not necessarily translate into unified theoretical frameworks or synthesized disciplinary architectures. There is no expected trajectory where every facet of reasoning in evo-devo dovetails in the long run, especially because we can be confident that problem agendas will alter and shift in their criteria of adequacy, take on new contours in their dimensions, and reconfigure in a variety of ways as disciplinary approaches grow, develop, fragment, and synthesize, sometimes in response to institutional pressures orthogonal to the research questions of evo-devo. However, this is not a reason for pessimism or despair. In fact, the opposite is true. The last several decades demonstrate that interdisciplinarity within evo-devo has yielded increasingly integrated explanatory frameworks for complex phenomena represented by structured problem agendas despite the existence of centrifugal forces on its disciplinary contributors and continued fragmentation of its epistemological outputs. And it is progress in our understanding of both the evolution of development and the developmental basis of evolution that is the ultimate goal, regardless of how our knowledge is structured or how the relevant social manifestations of the science are organized.

References

Adoutte, A., G. Balavoine, N. Lartillot, et al. 2000. The new animal phylogeny: Reliability and implications. *Proceedings of the National Academy of Sciences USA* 97:4453–4456.

Aguinaldo, A. M. A., J. M. Turbeville, L. S. Linford, et al. 1997. Evidence for a clade of nematodes, arthropods and other moulting animals. *Nature* 387:489–493.

Alberch, P. and E. A. Gale. 1985. A developmental analysis of an evolutionary trend: Digital reduction in amphibians. *Evolution* 39:8–23.

Albertson, R. C. 2018. Editorial. *Evolution & Development* 20:191.

Almudí, I., C. A. Martín-Blanco, I. M. García-Fernandez, et al. 2019. Establishment of the mayfly *Cloeon dipterum* as a new model system to investigate insect evolution. *EvoDevo* 10:6.

Almudí, I. and J. Pascual-Anaya. 2019. How do morphological novelties evolve? Novel approaches to define novel morphologies. In *Old Questions and Young Approaches to Animal Evolution*, edited by J. M. Martín-Durán and B. C. Vellutini. Cham: Springer, 107–132.

Amundson, R. 2005. *The Changing Role of the Embryo in Evolutionary Thought: Roots of Evo-Devo*. New York: Cambridge University Press.

Ankeny, R. and S. Leonelli. 2011. What's so special about model organisms? *Studies in History and Philosophy of Science* 42:313–323.

Antebi, A. 2013. Steroid regulation of *C. elegans* diapause, developmental timing, and longevity. In *Current Topics in Developmental Biology*, edited by E. R. Ann and B. O. C. Michael. San Diego: Academic Press, 181–212.

Arthur, W. 1984. *Mechanisms of Morphological Evolution: A Combined Genetic, Developmental and Ecological Approach*. Chichester: Wiley.

Arthur, W. 2002. The emerging conceptual framework of evolutionary developmental biology. *Nature* 415:757–764.

Arthur, W. 2004. *Biased Embryos and Evolution*. New York: Cambridge University Press.

Arthur, W. 2021. *Understanding Evo-Devo*. New York: Cambridge University Press.

Austin, C. J. 2017. Evo-devo: A science of dispositions. *European Journal for Philosophy of Science* 7(2):373–389.

Baedke, J. 2021. Mechanisms in evo-devo. In *Evolutionary Developmental Biology: A Reference Guide*, edited by L. Nuño de la Rosa and G. B. Müller. Cham: Springer International, 383–395.

Balari, S. and G. Lorenzo. 2021. Evo-devo of Language and Cognition. In *Evolutionary Developmental Biology: A Reference Guide*, edited by L. Nuno de la Rosa and G. Müller. Cham: Springer International Publishing 1221–1233.

Bier, E. and W. McGinnis. 2003. Model organisms in the study of development and disease. In *Molecular Basis of Inborn Errors of Development*, edited by C. J. Epstein, R. P. Erickson, and A. Wynshaw-Boris. New York: Oxford University Press, 25–45.

Bock, W. J. 2010. Multiple explanations in Darwinian evolutionary theory. *Acta Biotheoretica* 58:65–79.

Bolker, J. A. 1995. Model systems in developmental biology. *BioEssays* 17(5):451–455.

Bolker, J. A. 2009. Exemplary and surrogate models: Two modes of representation in biology. *Perspectives in Biology and Medicine* 52(4):485–499.

Bolker, J. A. 2014. Model species in evo-devo: A philosophical perspective. *Evolution & Development* 16:49–56.

Bonner, J. T., ed. 1982. *Evolution and Development*. Dahlem Conferenzen. Berlin: Springer-Verlag.

Brigandt, I. 2002. Homology and the origin of correspondence. *Biology & Philosophy* 17:389–407.

Brigandt, I. 2003. Homology in comparative, molecular, and evolutionary developmental biology. *Journal of Experimental Zoology (Mol Dev Evol)* 299B:9–17.

Brigandt, I. 2010. Beyond reduction and pluralism: Toward an epistemology of explanatory integration in biology. *Erkenntnis* 73:295–311.

Brigandt, I. 2015a. From developmental constraint to evolvability: How concepts figure in explanation and disciplinary identity. In *Conceptual Change in Biology: Scientific and Philosophical Perspectives on Evolution and Development*, edited by A. C. Love. Boston Studies in the Philosophy and History of Science. Dordrecht: Springer, 305–325.

Brigandt, I. 2015b. Evolutionary developmental biology and the limits of philosophical accounts of mechanistic explanation. In *Explanation in Biology: An Enquiry into the Diversity of Explanatory Patterns in the Life Sciences*, edited by P.-A. Braillard and C. Malaterre. Dordrecht: Springer, 135–173.

Brigandt, I. 2021. Typology and natural kinds in evo-devo. In *Evolutionary Developmental Biology: A Reference Guide*, edited by L. Nuño de la Rosa and G. B. Müller. Cham: Springer International, 483–493.

Brigandt, I. and A. C. Love. 2012. Conceptualizing evolutionary novelty: Moving beyond definitional debates. *Journal of Experimental Zoology (Mol Dev Evol)* 318B:417–427.

Brigandt, I., C. Villegas, A. C. Love, and L. Nuño de la Rosa. 2023. Evolvability as a disposition: Philosophical distinctions, scientific implications. In *Evolvability: A Unifying Concept in Evolutionary Biology?* edited by T. F. Hansen, D. Houle, M. Pavlicev, and C. Pélabon, Cambridge, MA: MIT Press, 55–72.

Brooks, D. S., J. DiFrisco, and W. C. Wimsatt, eds. 2021. *Levels of Organization in the Biological Sciences*. The Vienna Series in Theoretical Biology. Cambridge, MA: MIT Press.

Brown, R. L. 2014a. Identifying behavioral novelty. *Biological Theory* 9:135–148.

Brown, R. L. 2014b. What evolvability really is. *British Journal for the Philosophy of Science* 65(3):549–572.

Brown, R. L. 2021. Proximate versus ultimate causation and evo-devo. In *Evolutionary Developmental Biology: A Reference Guide*, edited by L. Nuño de la Rosa and G. B. Müller. Cham: Springer International, 425–433.

Burton, P. M. and J. R. Finnerty. 2009. Conserved and novel gene expression between regeneration and asexual fission in *Nematostella vectensis*. *Development Genes and Evolution* 219(2):79–87.

Calcott, B. 2009. Lineage explanations: Explaining how biological mechanisms change. *British Journal for Philosophy of Science* 60:51–78.

Caplan, A. L. 1978. Testability, disreputability, and the structure of the modern synthetic theory of evolution. *Erkenntnis* 13:261–278.

Carroll, S. B. 2005a. *Endless Forms Most Beautiful: The New Science of Evo-Devo*. New York: W. W. Norton.

Carroll, S. B. 2005b. Evolution at two levels: On genes and form. *PLOS Biology* 3(e245):1159–1166.

Carroll, S. B. 2008. Evo-devo and an expanding evolutionary synthesis: A genetic theory of morphological evolution. *Cell* 134:25–36.

Charbonneau, M. 2021. Evo-Devo and Culture. In *Evolutionary Developmental Biology: A Reference Guide*, edited by L. Nuno de la Rosa and G. Müller. Cham: Springer International Publishing 1235–1248.

Cheon, D.-J. and S. Orsulic. 2011. Mouse models of cancer. *Annual Review of Pathology: Mechanisms of Disease* 6(1):95–119.

Chipman, A. D. 2015. Hexapoda: Comparative aspects of early development. In *Evolutionary Developmental Biology of Invertebrates 5: Ecdysozoa III: Hexapoda*, edited by A. Wanninger. Vienna: Springer, 93–110.

Collins, J. P. 2002. May you live in interesting times: Using multidisciplinary and interdisciplinary programs to cope with change in the life sciences. *BioScience* 52(1):75–83.

Craver, C. F. and L. Darden. 2013. *In Search of Mechanisms: Discoveries across the Life Sciences*. Chicago: University of Chicago Press.

Craver, C. F. and J. Tabery. 2016. Mechanisms in science. In *The Stanford Encyclopedia of Philosophy*, edited by E. N. Zalta, http://plato.stanford.edu/archives/spr2016/entries/science-mechanisms/.

Damen, W. G. M. 2007. Evolutionary conservation and divergence of the segmentation process in arthropods. *Developmental Dynamics* 236: 1379–1391.

Darling, J. A., A. R. Reitzel, P. M. Burton, et al. 2005. Rising starlet: The starlet sea anemone, *Nematostella vectensis*. *BioEssays* 27(2):211–221.

Darwin, C. 1964. *On the Origin of Species: A Facsimile of the First Edition*. Cambridge, MA: Harvard University Press.

Davidson, E. H. 2006. *The Regulatory Genome: Gene Regulatory Networks in Development and Evolution*. San Diego: Academic Press.

Davidson, E. H. and D. H. Erwin. 2006. Gene regulatory networks and the evolution of animal body plans. *Science* 311:796–800.

Davidson, E. H. and I. S. Peter. 2015. *Genomic Control Process: Development and Evolution*. San Diego: Academic Press.

Davis, G. K. and N. H. Patel. 1999. The origin and evolution of segmentation. *Trends in Genetics* 9:M68–M72.

De Robertis, E. M. 2008. Evo-devo: Variations on ancestral themes. *Cell* 132:185–195.

Depew, D. J. and B. H. Weber. 1996. *Darwinism Evolving: Systems Dynamics and the Genealogy of Natural Selection*. Cambridge, MA: MIT Press.

Di Stilio, V. S. and S. M. Ickert-Bond. 2021. *Ephedra* as a gymnosperm evo-devo model lineage. *Evolution & Development* 23(3):256–266.

Dickins, T. E. and Q. Rahman. 2012. The extended evolutionary synthesis and the role of soft inheritance in evolution. *Proceedings of the Royal Society B: Biological Sciences* 279:2913–2921.

Dietrich, M. R., R. A. Ankeny, N. Crowe, S. Green, and S. Leonelli. 2020. How to choose your research organism. *Studies in History and Philosophy of Biological and Biomedical Sciences* 80:101227.

DiFrisco, J. 2023. Toward a theory of homology: Development and the de-coupling of morphological and molecular evolution. *British Journal for the Philosophy of Science* 74:771–810.

DiFrisco, J., A. C. Love, and G. P. Wagner. 2020. Character identity mechanisms: A conceptual model for comparative-mechanistic biology. *Biology & Philosophy* 35(4):44.

DiFrisco, J., A. C. Love, and G. P. Wagner. 2023a. The hierarchical basis of serial homology and evolutionary novelty. *Journal of Morphology* 284(1): e21531.

DiFrisco, J., G. P. Wagner, and A. C. Love. 2023b. Reframing research on evolutionary novelty and co-option: Character identity mechanisms versus deep homology. *Seminars in Cell &Developmental Biology* 145:3–12.

DiTeresi, C. A. 2010. *Taming Variation: Typological Thinking and Scientific Practice in Developmental Biology.* PhD dissertation, Committee on Conceptual and Historical Studies of Science, University of Chicago.

Dupré, J. 2001. In defence of classification. *Studies the History and Philosophy of Biological and Biomedical Sciences* 32(2):203–219.

Elliott, S. 2021. Research problems. *British Journal for the Philosophy of Science* 72:1013–1037.

Erwin, D. H. 2012. Novelties that change carrying capacity. *Journal of Experimental Zoology (Mol Dev Evol)* 318(6):460–465.

Erwin, D. H. 2020. The origin of animal body plans: A view from fossil evidence and the regulatory genome. *Development* 147(4):dev182899.

Erwin, D. H. 2021. A conceptual framework of evolutionary novelty and innovation. *Biological Reviews* 96(1):1–15.

Fábregas-Tejeda, A. and F. Vergara-Silva. 2018. The emerging structure of the extended evolutionary synthesis: Where does evo-devo fit in? *Theory in Biosciences* 137(2):169–184.

Feder, M. E. and T. Mitchell-Olds. 2003. Evolutionary and ecological functional genomics. *Nature Reviews Genetics* 4:649–655.

Field, K. G., G. J. Olsen, D. J. Lane, et al. 1988. Molecular phylogeny of the animal kingdom. *Science* 239:748–753.

Frankino, W. A. and R. A. Raff. 2004. Evolutionary importance and pattern of phenotypic plasticity. In *Phenotypic Plasticity: Functional and Conceptual Approaches*, edited by T. J. DeWitt and S. M. Scheiner. New York: Oxford University Press, 64–81.

Fraser, S. E. and R. M. Harland. 2000. The molecular metamorphosis of experimental embryology. *Cell* 100:41–55.

Futuyma, D. J. 2017. Evolutionary biology today and the call for an extended synthesis. *Interface Focus* 7(5):20160145.

Garstang, W. 1928. The morphology of the Tunicata, and its bearings on the phylogeny of the Chordata. *Quarterly Journal of Microscopical Science* 72:51–187.

Gayon, J. 1990. Critics and criticisms of the Modern Synthesis: The viewpoint of a philosopher. In *Evolutionary Biology, Volume 24*, edited by M. K. Hecht, B. Wallace, and R. J. Macintyre. New York: Plenum Press, 1–49.

Genikhovich, G. and U. Technau. 2009. The starlet sea anemone *Nematostella vectensis*: An anthozoan model organism for studies in comparative genomics and functional evolutionary developmental biology. *Cold Spring Harbor Protocols* 4(9):pdb.emo129.

Ghiselin, M. T. 2005. Homology as a relation of correspondence between parts of individuals. *Theory in Biosciences* 124:91–103.

Gilbert, S. F. 1997. *Developmental Biology*. Sunderland, MA: Sinauer Associates.

Gilbert, S. F. and R. M. Burian. 2003. Development, evolution, and evolutionary developmental biology. In *Keywords and Concepts in Evolutionary Developmental Biology*, edited by B. K. Hall and W. M. Olson. Cambridge, MA: Harvard University Press, 61–68.

Gilbert, S. F. and D. Epel. 2009. *Ecological Developmental Biology: Integrating Epigenetics, Medicine, and Evolution*. Sunderland, MA: Sinauer.

Giribet, G. and G. D. Edgecombe. 2020. *The Invertebrate Tree of Life*. Princeton: Princeton University Press.

Glennan, S., P. Illari, and E. Weber. 2022. Six theses on mechanisms and mechanistic science. *Journal for General Philosophy of Science* 53(2):143–161.

Gompel, N. and B. Prud'homme. 2009. The causes of repeated genetic evolution. *Developmental Biology* 332:36–47.

Gompel, N., B. Prud'homme, P. J. Wittkopp, V. A. Kassner, and S. B. Carroll. 2005. Chance caught on the wing: *cis*-regulatory evolution and the origin of pigment patterns in *Drosophila*. *Nature* 433:481–487.

Goodwin, B. C., N. Holder, and C. C. Wylie, eds. 1983. *Development and Evolution*. Cambridge: Cambridge University Press.

Gould, S. J. 1977. *Ontogeny and Phylogeny*. Cambridge, MA: Belknap/Harvard University Press.

Gouvea, D. S. Y. 2020. *Essentially Dynamic Concepts and the Case of Homology*. PhD Dissertation, Committee on Conceptual and Historical Studies of Science, University of Chicago.

Green, S., M. R. Dietrich, S. Leonelli, and R. A. Ankeny. 2018. "Extreme" organisms and the problem of generalization: Interpreting the Krogh principle. *History and Philosophy of the Life Sciences* 40:65.

Griesemer, J. 1984. Presentations and the status of theories. *PSA: Proceedings of the Biennial Meeting of the Philosophy of Science Association*. Volume One, Contributed Papers, 102–114.

Griesemer, J. 2015. What salamander biologists have taught us about evo-devo. In *Conceptual Change in Biology: Scientific and Philosophical Perspectives on Evolution and Development*, edited by A. C. Love. Boston Studies in the Philosophy and History of Science. Dordrecht: Springer, 271–301.

Haag, E. S. 2014. The same but different: Worms reveal the pervasiveness of developmental system drift. *PLoS Genetics* 10(2):e1004150.

Halanych, K. M., J. D. Bacheller, A. M. Aguinaldo, et al. 1995. Evidence from 18S ribosomal DNA that the lophophorates are protostome animals. *Science* 267:1641–1643.

Hall, B. K. 1999. *Evolutionary Developmental Biology*. Dordrecht: Kluwer Academic.

Hall, B. K. 2002. Palaeontology and evolutionary developmental biology: A science of the nineteenth and twenty-first centuries. *Palaeontology* 45(4):647–669.

Hall, B. K., ed. 2007. *Fins into Limbs: Evolution, Development, and Transformation*. Chicago: University of Chicago Press.

Hall, B. K. and B. Hallgrímsson. 2008. *Strickberger's Evolution*. 4th ed. Sudbury, MA: Jones and Bartlett.

Hall, B. K. and R. Kerney. 2012. Levels of biological organization and the origin of novelty. *Journal of Experimental Zoology (Mol Dev Evol)* 318B:428–437.

Hancock, Z. B., E. S. Lehmberg, and G. S. Bradburd. 2021. Neo-Darwinism still haunts evolutionary theory: A modern perspective on Charlesworth, Lande, and Slatkin (1982). *Evolution* 75(6):1244–1255.

Hansen, T. F. and D. Houle. 2008. Measuring and comparing evolvability and constraint in multivariate characters. *Journal of Evolutionary Biology* 21(5):1201–1219.

Hansen, T. F., D. Houle, M. Pavlicev, and C. Pélabon, eds. 2023. *Evolvability: A Unifying Concept in Evolutionary Biology?* Cambridge, MA: MIT Press.

Hansson, B. 1999. Interdisciplinarity: For what purpose? *Policy Sciences* 32:339–343.

Hattiangadi, J. N. 1978. The structure of problems, part I. *Philosophy of the Social Sciences* 8:345–365.

Hattiangadi, J. N. 1979. The structure of problems, part II. *Philosophy of the Social Sciences* 9:49–76.

He, S., F. del Viso, C.-Y. Chen, A. Ikmi, A.E. Kroesen and M.C. Gibson. 2018. An axial Hox code controls tissue segmentation and body patterning in Nematostella vectensis. *Science* 361:1377–1380.

Held, L. I. 2017. *Deep Homology? Uncanny Similarities of Humans and Flies Uncovered by Evo-devo*. New York: Cambridge University Press.

Hendrikse, J. L., T. E. Parsons, and B. Hallgrímsson. 2007. Evolvability as the proper focus of evolutionary developmental biology. *Evolution & Development* 9(4):393–401.

Hoekstra, H. E. and J. A. Coyne. 2007. The locus of evolution: Evo-devo and the genetics of adaptation. *Evolution* 61(5):995–1016.

Holland, L. Z., P. W. H. Holland, and N. D. Holland. 1996. Revealing homologies between body parts of distantly related animals by *in situ* hybridization to developmental genes: Amphioxus versus vertebrates. In *Molecular Zoology: Advances, Strategies, and Protocols*, edited by J. D. Ferraris and S. R. Palumbi. New York: Wiley-Liss, 267–282.

Holland, P. W. H. 1999. The future of evolutionary developmental biology. *Nature* 402:C41–C44.

Hopwood, N. 2007. A history of normal plates, tables and stages in vertebrate embryology. *International Journal of Developmental Biology* 51:1–26.

Houle, D. 1992. Comparing evolvability and variability of quantitative traits. *Genetics* 130:195–204.

Hughes, C. L. and T. C. Kaufman. 2002. *Hox* genes and the evolution of the arthropod body plan. *Evolution & Development* 4(6):459–499.

Hüttemann, A. and M. I. Kaiser. 2018. Potentiality in biology. In *Handbook of Potentiality*, edited by K. Engelhard and M. Quante. Dordrecht: Springer, 401–428.

Illari, P. and J. Williamson. 2012. What is a mechanism? Thinking about mechanisms across the sciences. *European Journal of the Philosophy of Science* 2:119–135.

Jablonski, D. 1987. Heritability at the species level: Analysis of geographic ranges of Cretaceous mollusks. *Science* 238:360–363.

Janssen, R., M. Le Gouar, M. Pechmann, et al. 2010. Conservation, loss, and redeployment of Wnt ligands in protostomes: Implications for understanding the evolution of segment formation. *BMC Evolutionary Biology* 10(1):1–21.

Jenner, R. A. 2000. Evolution of animal body plans: The role of metazoan phylogeny at the interface between pattern and process. *Evolution & Development* 2(4):208–221.

Jenner, R. A. 2006. Unburdening evo-devo: Ancestral attractions, model organisms, and basal baloney. *Development Genes and Evolution* 216:385–394.

Jenner, R. A. 2022. *Ancestors in Evolutionary Biology: Linear Thinking about Branching Trees*. Cambridge: Cambridge University Press.

Jenner, R. A. and M. A. Wills. 2007. The choice of model organisms in evo-devo. *Nature Reviews Genetics* 8:311–319.

Jones, M. R. 2005. Idealization and abstraction: A framework. In *Idealization XII: Correcting the Model: Idealization and Abstraction in the Sciences*, edited by M. R. Jones and N. Cartwright. Amsterdam: Rodopi, 173–217.

Kawano, S. M. and R. W. Blob. 2013. Propulsive forces of mudskipper fins and salamander limbs during terrestrial locomotion: Implications for the invasion of land. *Integrative & Comparative Biology* 53(2):283–294.

Kellert, S. H., H. E. Longino, and C. K. Waters. 2006. Introduction: The pluralist stance. In *Scientific Pluralism*, edited by S. H. Kellert, H. E. Longino, and C. K. Waters. Minnesota Studies in Philosophy of Science. Minneapolis: University of Minnesota Press, vii–xxix.

Kimmel, C. B., W. W. Ballard, S. R. Kimmel, B. Ullmann, and T. F. Schilling. 1995. Stages of embryonic development of the zebrafish. *Developmental Dynamics* 203(3):253–310.

Kirschner, M. W. 2015. The road to facilitated variation. In *Conceptual Change in Biology: Scientific and Philosophical Perspectives on Evolution and Development*, edited by A. C. Love. Boston Studies in the Philosophy and History of Science. Dordrecht: Springer, 199–217.

Kirschner, M. W. and J. C. Gerhart. 1998. Evolvability. *Proceedings of the National Academy of Sciences USA* 95:8420–8427.

Kirschner, M. W. and J. C. Gerhart. 2005. *The Plausibility of Life: Resolving Darwin's Dilemma*. New Haven: Yale University Press.

Krogh, A. 1929. The progress of physiology. *Science* 70:200–204.

Kuo, D.-H. and Y.-T. Lai. 2019. On the origin of leeches by evolution of development. *Development, Growth & Differentiation* 61(1):43–57.

Kuratani, S., S. Kuraku, and H. Nagashima. 2011. Evolutionary developmental perspective for the origin of turtles: The folding theory for the shell based on the developmental nature of the carapacial ridge. *Evolution & Development* 13(1):1–14.

Kusserow, A., K. Pang, C. Sturm, et al. 2005. Unexpected complexity of the *Wnt* gene family in a sea anemone. *Nature* 433:156–160.

Laland, K. N., T. Uller, M. W. Feldman, et al. 2015. The extended evolutionary synthesis: Its structure, assumptions and predictions. *Proceedings of the Royal Society B: Biological Sciences* 282:20151019.

Lapraz, F., K. A. Rawlinson, J. Girstmair, et al. 2013. Put a tiger in your tank: The polyclad flatworm *Maritigrella crozieri* as a proposed model for evo-devo. *EvoDevo* 4(1):29.

Larsson, H. C. E. and G. P. Wagner. 2012. Testing inferences in developmental evolution: The forensic evidence principle. *Journal of Experimental Zoology (Mol Dev Evol)* 318B:489–500.

Laubichler, M. D. 2009. Form and function in Evo Devo: Historical and conceptual reflections. In *Form and Function in Developmental Evolution*, edited by M. D. Laubichler and J. Maienschein. New York: Cambridge University Press, 10–46.

Laubichler, M. D. 2010. Evolutionary developmental biology offers a significant challenge to the neo-Darwinian paradigm. In *Contemporary Debates in Philosophy of Biology*, edited by F. J. Ayala and R. Arp. Malden, MA: Wiley-Blackwell, 199–212.

Laubichler, M. D. and J. Maienschein, eds. 2007. *From Embryology to Evo-devo: A History of Developmental Evolution*. Cambridge, MA: MIT Press.

Laudan, L. 1977. *Progress and Its Problems: Towards a Theory of Scientific Growth*. Berkeley: University of California Press.

Layden, M. J., F. Rentzsch, and E. Röttinger. 2016. The rise of the starlet sea anemone *Nematostella vectensis* as a model system to investigate development and regeneration. *Wiley Interdisciplinary Reviews: Developmental Biology* 5(4):408–428.

Lev, O. and A. D. Chipman. 2021. Development of the pre-gnathal segments in the milkweed bug *Oncopeltus fasciatus* suggests they are not serial homologs of trunk segments. *Frontiers in Cell and Developmental Biology* 9:695135.

Levis, N. A. and D. W. Pfennig. 2019. Phenotypic plasticity, canalization, and the origins of novelty: Evidence and mechanisms from amphibians. *Seminars in Cell & Developmental Biology* 88:80–90.

Lloyd, E. A. 1988. *The Structure and Confirmation of Evolutionary Theory*. Westport, CT: Greenwood Press.

Love, A. C. 2003. Evolvability, dispositions, and intrinsicality. *Philosophy of Science* 70(5):1015–1027.

Love, A. C. 2006. Evolutionary morphology and evo-devo: Hierarchy and novelty. *Theory in Biosciences* 124:317–333.

Love, A. C. 2007a. Functional homology and homology of function: Biological concepts and philosophical consequences. *Biology & Philosophy* 22:691–708.

Love, A. C. 2007b. Morphological and paleontological perspectives for a history of evo-devo. In *From Embryology to Evo-Devo: A History of Developmental Evolution*, edited by M. Laubichler and J. Maienschein. Cambridge, MA: MIT Press, 267–307.

Love, A. C. 2008. Explaining evolutionary innovation and novelty: Criteria of explanatory adequacy and epistemological prerequisites. *Philosophy of Science* 75:874–886.

Love, A. C. 2009a. Marine invertebrates, model organisms, and the Modern Synthesis: Epistemic values, evo-devo, and exclusion. *Theory in Biosciences* 128:19–42.

Love, A. C. 2009b. Typology reconfigured: From the metaphysics of essentialism to the epistemology of representation. *Acta Biotheoretica* 57:51–75.

Love, A. C. 2010a. Rethinking the structure of evolutionary theory for an extended synthesis. In *Evolution – The Extended Synthesis*, edited by M. Pigliucci and G. B. Müller. Cambridge, MA: MIT Press, 403–441.

Love, A. C. 2010b. Idealization in evolutionary developmental investigation: A tension between phenotypic plasticity and normal stages. *Philosophical Transactions of the Royal Society B: Biological Sciences* 365:679–690.

Love, A. C. 2013. Theory is as theory does: Scientific practice and theory structure in biology. *Biological Theory* 7:325–337.

Love, A. C. 2015a. Conceptual change and evolutionary developmental biology. In *Conceptual Change in Biology: Scientific and Philosophical Perspectives on Evolution and Development*, edited by A. C. Love. Boston Studies in the Philosophy and History of Science. Dordrecht: Springer, 1–53.

Love, A. C. 2015b. Evolutionary developmental biology: Philosophical issues. In *Handbook of Evolutionary Thinking in the Sciences*, edited by T. Heams, P. Huneman, L. Lecointre, and M. Silberstein. Berlin: Springer, 265–283.

Love, A. C. 2018. Developmental mechanisms. In *The Routledge Handbook of the Philosophy of Mechanisms and Mechanical Philosophy*, edited by G. S. and P. Illari. New York: Routledge, 332–347.

Love, A. C. 2020. Situating evolutionary developmental biology in evolutionary theory. In *The Theory of Evolution: Principles, Concepts, and Assumptions*, edited by S. M. Scheiner and D. P. Mindell. Chicago: University of Chicago Press, 144–168.

Love, A. C. 2021. Interdisciplinarity in evo-devo. In *Evolutionary Developmental Biology: A Reference Guide*, edited by L. Nuño de la Rosa and G. B. Müller. Cham: Springer International, 407–423.

Love, A. C., M. Grabowski, D. Houle, et al. 2022. Evolvability in the fossil record. *Paleobiology* 48(2):186–209.

Love, A. C. and G. L. Lugar. 2013. Dimensions of integration in interdisciplinary explanations of the origin of evolutionary novelty. *Studies in the History and Philosophy of Biological and Biomedical Sciences* 44:537–550.

Love, A. C. and R. A. Raff. 2003. Knowing your ancestors: Themes in the history of evo-devo. *Evolution & Development* 5(4):327–330.

Love, A. C., T. A. Stewart, G. P. Wagner and S. A. Newman. 2017. Perspectives on integrating genetic and physical explanations of evolution and development. *Integrative & Comparative Biology* 57:1258–1268.

Love, A. C. and R. R. Strathmann. 2018. Marine invertebrate larvae: Model life histories for development, ecology, and evolution. In *Evolutionary Ecology of Marine Invertebrate Larvae*, edited by T. J. Carrier, A. M. Reitzel, and A. Heyland. Oxford: Oxford University Press, 302–317.

Love, A. C. and M. Travisano. 2013. Microbes modeling ontogeny. *Biology & Philosophy* 28:161–188.

Love, A. C. and G. P. Wagner. 2022. Co-option of stress mechanisms in the origin of evolutionary novelties. *Evolution* 76(3):394–413.

Love, A. C. and Y. Yoshida. 2019. Reflections on model organisms in evolutionary developmental biology. In *Evo-Devo: Non-model Species in Cell and Developmental Biology*, edited by W. Tworzydlo and S. M. Bilinski. Cham: Springer, 3–20.

Lowe, J. W. E. 2015. Managing variation in the investigation of organismal development: Problems and opportunities. *History and Philosophy of the Life Sciences* 37:449–473.

Lowe, J. W. E. 2016. Normal development and experimental embryology: Edmund Beecher Wilson and Amphioxus. *Studies in History and Philosophy of Biological and Biomedical Sciences* 57:44–59.

Lowe, C. J., D. N. Clarke, D. M. Medeiros, D. S. Rokhsar, and J. Gerhart. 2015. The deuterostome context of chordate origins. *Nature* 520:456–465.

Lum, L. and P. A. Beachy. 2004. The Hedgehog response network: Sensors, switches, and routers. *Science* 304:1755–1759.

Lustig, A. J. 2009. Darwin's difficulties. In *The Cambridge Companion to the "Origin of Species,"* edited by M. Ruse and R. J. Richards. New York: Cambridge University Press, 109–128.

Lynch, M. 2007. The frailty of adaptive hypotheses for the origins of organismal complexity. *Proceedings of the National Academy of Sciences USA* 104:8597–8604.

Lynch, V. J. 2009. Use with caution: Developmental systems divergence and potential pitfalls of animal models. *Yale Journal of Biology and Medicine* 82:53–66.

Lynch, V. J., G. May, and G. P. Wagner. 2011. Regulatory evolution through divergence of a phosphoswitch in the transcription factor CEBPB. *Nature* 480:383–386.

Lynch, V. J. and G. P. Wagner. 2008. Resurrecting the role of transcription factor change in developmental evolution. *Evolution* 62(9):2131–2154.

Lyson, T. R., G. S. Bever, T. M. Scheyer, A. Y. Hsiang, and J. A. Gauthier. 2013. Evolutionary origin of the turtle shell. *Current Biology* 23(12): 1113–1119.

Mabee, P. M., K. L. Olmstead, and C. C. Cubbage. 2000. An experimental study of intraspecific variation, developmental timing, and heterochrony in fishes. *Evolution* 54(6):2091–2106.

Maynard Smith, J., R. Burian, S. Kauffman, et al. 1985. Developmental constraints and evolution. *Quarterly Review of Biology* 60(3):265–287.

Mayr, E. 1960. The emergence of evolutionary novelties. In *Evolution after Darwin. Volume 1: The Evolution of Life, Its Origin, History and Future*, edited by S. Tax. Chicago: University of Chicago Press, 349–380.

McDonnell, M. A. 2016. The American War for Independence as a Revolutionary War. In *Oxford Research Encyclopedia of American History*, edited by J. Butler : Oxford University Press: http://oxfordre.com/americanhistory/view/10.1093/acrefore/9780199329175.9780199329001.9780199320001/acrefore-9780199329175-e-9780199329171.

McGinnis, W., R. L. Garber, J. Wirz, A. Kuroiwa, and W. J. Gehring. 1984. A homologous protein-coding sequence in *Drosophila* homeotic genes and its conservation in other metazoans. *Cell* 37:403–408.

McGinnis, W. and R. Krumlauf. 1992. Homeobox genes and axial patterning. *Cell* 68(2):283–302.

Milinkovitch, M. and A. Tzika. 2007. Escaping the mouse trap: The selection of new evo-devo species. *Journal of Experimental Zoology (Mol Dev Evol)* 308B:337–346.

Minelli, A. 2003. *The Development of Animal Form: Ontogeny, Morphology, and Evolution*. Cambridge: Cambridge University Press.

Minelli, A. and J. Baedke. 2014. Model organisms in evo-devo: Promises and pitfalls of the comparative approach. *History and Philosophy of the Life Sciences* 36(1):42–59.

Minelli, A., C. Brena, G. Deflorian, D. Maruzzo, and G. Fusco. 2006. From embryo to adult-beyond the conventional periodization of arthropod development. *Development Genes and Evolution* 216:373–383.

Minelli, A. and G. Fusco. 2004. Evo-devo perspectives on segmentation: Model organisms, and beyond. *Trends in Ecology & Evolution* 19(8):423–429.

Mitteröcker, P. 2021. Morphometrics in evolutionary developmental biology. In *Evolutionary Developmental Biology: A Reference Guide*, edited by L. Nuño de la Rosa and G. B. Müller. Cham: Springer, 941–951.

Moczek, A. 2007. Developmental capacitance, genetic accommodation, and adaptive evolution. *Evolution & Development* 9:299–305.

Moczek, A. 2008. On the origins of novelty in development and evolution. *BioEssays* 30(5):432–447.

Moczek, A. P., K. E. Sears, A. Stollewerk, et al. 2015. The significance and scope of evolutionary developmental biology: A vision for the 21st century. *Evolution & Development* 17:198–219.

Moczek, A. P., S. Sultan, S. Foster, et al. 2011. The role of developmental plasticity in evolutionary innovation. *Proceedings of the Royal Society of London B: Biological Sciences* 278:2705–2713.

Müller, G. B. 2007. Evo-devo: Extending the evolutionary synthesis. *Nature Reviews Genetics* 8:943–949.

Müller, G. B. 2010. Epigenetic innovation. In *Evolution – The Extended Synthesis*, edited by M. Pigliucci and G. B. Müller. Cambridge, MA: MIT Press, 307–332.

Müller, G. B. 2017. Why an extended evolutionary synthesis is necessary. *Interface Focus* 7(5):20170015.

Müller, G. B. and G. P. Wagner. 1991. Novelty in evolution: Restructuring the concept. *Annual Review of Ecology and Systematics* 22:229–256.

Nakamura, T., A. R. Gehrke, J. Lemberg, J. Szymaszek, and N. H. Shubin. 2016. Digits and fin rays share common developmental histories. *Nature* 537:225–228.

National Academy of Sciences. 2005. *Facilitating Interdisciplinary Research*. Washington, DC: National Academies Press.

Newman, S. A. 2012. Physico-genetic determinants in the evolution of development. *Science* 338:217–219.

Newman, S. A., G. Forgacs, and G. B. Müller. 2006. Before programs: The physical origination of multicellular forms. *International Journal of Developmental Biology* 50:289–299.

Newman, S. A. and G .B. Müller. 2005. Origination and innovation in the vertebrate limb skeleton: An epigenetic perspective. *Journal of Experimental Zoology (Mol Dev Evol)* 304B(6):593–609.

Nickles, T. 1981. What is a problem that we may solve it? *Synthese* 47:85–118.

Novick, A. 2018. The fine structure of "homology." *Biology & Philosophy* 33:6.

Novick, A. 2019. Cuvierian functionalism. *Philosophy, Theory, & Practice in Biology* 11(5).

Nuño de la Rosa, L. 2017. Computing the extended synthesis: Mapping the dynamics and conceptual structure of the evolvability research front. *Journal of Experimental Zoology (Mol Dev Evol)* 328B:395–411.

Nuño de la Rosa, L and G. B. Müller, eds. 2021. *Evolutionary Developmental Biology: A Reference Guide*. Cham: Springer.

Nunes, M. D. S., S. Arif, C. Schlötterer, and A. P. McGregor. 2013. A perspective on micro-evo-devo: Progress and potential. *Genetics* 195(3):625–634.

Oakley, T. H. and D. I. Speiser. 2015. How complexity originates: The evolution of animal eyes. *Annual Review of Ecology, Evolution, and Systematics* 46(1):237–260.

Olsson, L. and B. K. Hall. 1999. Developmental and evolutionary perspectives on major transformations in body organization. *American Zoologist* 39:612–616.

Onimaru, K., L. Marcon, M. Musy, M. Tanaka, and J. Sharpe. 2016. The fin-to-limb transition as the re-organization of a Turing pattern. *Nature Communications* 7(1):11582.

Oster, G. F., N. Shubin, J. D. Murray, and P. Alberch. 1988. Evolution and morphogenetic rules: The shape of the vertebrate limb in ontogeny and phylogeny. *Evolution* 42:862–884.

Owen, R. 1843. *Lectures on the Comparative Anatomy and Physiology of the Invertebrate Animals*. London: Longman, Brown, Green, and Longmans.

Palmer, A. R. 2012. Developmental plasticity and the origin of novel forms: Unveiling cryptic genetic variation via "use and disuse." *Journal of Experimental Zoology (Mol Dev Evol)* 318B:466–479.

Pavlicev, M. and G. P. Wagner. 2012. A model of developmental evolution: Selection, pleiotropy and compensation. *Trends in Ecology & Evolution* 27(6):316–322.

Pavlicev, M. and S. Widder. 2015. Wiring for independence: Positive feedback motifs facilitate individuation of traits in development and evolution. *Journal of Experimental Zoology (Mol Dev Evol)* 324(2):104–113.

Petersen, C. P. and P. W. Reddien. 2009. Wnt signaling and the polarity of the primary body axis. *Cell* 139(6):1056–1068.

Piatigorsky, J. 2007. *Gene Sharing and Evolution: The Diversity of Protein Functions*. Cambridge, MA: Harvard University Press.

Pigliucci, M. 2007. Do we need an extended evolutionary synthesis? *Evolution* 61:2743–2749.

Pigliucci, M. and G. B. Müller. 2010. Elements of an extended evolutionary synthesis. In *Evolution – The Extended Synthesis*, edited by M. Pigliucci and G. B. Müller. Cambridge, MA: MIT Press, 3–17.

Ploeger, A. and F. Galis. 2021. Evo-devo and Cognitive Science. In *Evolutionary Developmental Biology: A Reference Guide*, edited by L. Nuno de la Rosa and G. Müller. Cham: Springer International Publishing 1209–1220.

Popper, K. 2002 [1963]. *Conjectures and Refutations: The Growth of Scientific Knowledge*. LondonRoutledge.

Quiring, R., U. Walldorf, U. Kloter, and W. J. Gehring. 1994. Homology of the *eyeless* gene of *Drosophila* to the *Small eye* gene in mice and *Aniridia* in humans. *Science* 265:785–789.

Raff, R. A. 2000. Evo-devo: The evolution of a new discipline. *Nature Reviews Genetics* 1:74–79.

Raff, R. A. 2007. Written in stone: Fossils, genes, and evo-devo. *Nature Reviews Genetics* 8:911–920.

Raff, R. A. and T. C. Kaufman. 1983. *Embryos, Genes, and Evolution: The Developmental-Genetic Basis of Evolutionary Change*. New York: Macmillan.

Reed, R. D., P.-H. Chen, and H. F. Nijhout. 2007. Cryptic variation in butterfly eyespot development: The importance of sample size in gene expression studies. *Evolution & Development* 9:2–9.

Reiss, J. O. 2003. Time. In *Keywords and Concepts in Evolutionary Developmental Biology*, edited by B. K. Hall and W. M. Olson. Cambridge, MA: Harvard University Press, 359–368.

Remane, A. 1952. *Die Grundlagen des natürlichen Systems, der vergleichenden Anatomie und der Phylogenetik*. Leipzig: Akademische Verlagsgesellschaft Geest & Portig K. G.

Repko, A. F. 2008. *Interdisciplinary Research: Process and Theory*. Thousand Oaks, CA: Sage.

Rice, S. H. 2004. *Evolutionary Theory: Mathematical and Conceptual Foundations*. Sunderland, MA: Sinauer Associates.

Rice, S. H. 2012. The place of development in mathematical evolutionary theory. *Journal of Experimental Zoology (Mol Dev Evol)* 318B: 480–488.

Richardson, M. K. 2022. Theories, laws, and models in evo-devo. *Journal of Experimental Zoology (Mol Dev Evol)* 338B:36–61.

Riedl, R. 1978. *Order in Living Organisms: A Systems Analysis of Evolution*. New York: John Wiley.

Rohner, P. T., A. L. M. Macagno, and A. P. Moczek. 2020. Evolution and plasticity of morph-specific integration in the bull-headed dung beetle *Onthophagus taurus*. *Ecology and Evolution* 10:10558–10570.

Röttinger, E., P. Dahlin, and M. Q. Martindale. 2012. A framework for the establishment of a Cnidarian gene regulatory network for "endomesoderm" specification: The inputs of ß-Catenin/TCF Signaling. *PLOS Genetics* 8(12): e1003164.

Russell, J. J., J. A. Theriot, P. Sood, et al. 2017. Non-model model organisms. *BMC Biology* 15(1):55.

Salazar-Ciudad, I. and H. Cano-Fernández. 2023. Evo-devo beyond development: Generalizing evo-devo to all levels of the phenotypic evolution. *BioEssays* 45(3):2200205.

Salazar-Ciudad, I., M. Marín-Riera, and M. Brun-Usan. 2021. Understanding the genotype-phenotype map: Contrasting mathematical models. In *Evolutionary Systems Biology: Advances, Questions, and Opportunities*, edited by A. Crombach. Cham: Springer, 221–244.

Sanger, T. J. 2012. The emergence of squamates as model systems for integrative biology. *Evolution & Development* 14(3):231–233.

Scheiner, S. M. 2010. Toward a conceptual framework for biology. *The Quarterly Review of Biology* 85:293–318.

Schierwater, B., M. Eitel, W. Jakob, et al. 2009. Concatenated analysis sheds light on early metazoan evolution and fuels a modern "Urmetazoon" hypothesis. *PLOS Biology* 7(1):e1000020.

Scotland, R. W. 2010. Deep homology: A view from systematics. *BioEssays* 32(5):438–449.

Scott, M. P. and A. J. Weiner. 1984. Structural relationships among genes that control development: Sequence homology between the *Antennapedia*, *Ultrabithorax*, and *fushi tarazu* loci of *Drosophila*. *Proceedings of the National Academy of Sciences USA* 81(13):4115–4119.

Sheth, R., L. Marcon, M. F. l. Bastida, et al. 2012. *Hox* genes regulate digit patterning by controlling the wavelength of a Turing-type mechanism. *Science* 338:1476–1480.

Shigetani, Y., F. Sugahara, and S. Kuratani. 2005. A new evolutionary scenario for the vertebrate jaw. *BioEssays* 27(3):331–338.

Shirai, L. T., S. V. Saenko, R. A. Keller, et al. 2012. Evolutionary history of the recruitment of conserved developmental genes in association to the formation and diversification of a novel trait. *BMC Evolutionary Biology* 21:21.

Shubin, N. H., C. Tabin, and S. B. Carroll. 2009. Deep homology and the origins of evolutionary novelty. *Nature* 457:818–823.

Shubin, N. H., C. Tabin, and S. B. Carroll. 1997. Fossils, genes and the evolution of animal limbs. *Nature* 388:639–648.

Shubin, N. H., E. B. Daeschler, and F. A. Jenkins. 2006. The pectoral fin of *Tiktaalik roseae* and the origin of the tetrapod limb. *Nature* 440:764–771.

Slack, J. M. W. 2006. *Essential Developmental Biology*. Malden, MA: Blackwell.

Sommer, R. J. 2009. The future of evo-devo: Model systems and evolutionary theory. *Nature Reviews Genetics* 10(6):416–422.

Sterelny, K. 2007. What is evolvability? In *Philosophy of Biology (Handbook of Philosophy of Science)*, edited by M. Matthen and C. Stephens. Amsterdam: Elsevier, 163–178.

Stewart, T. A., R. Bhat, and S. A. Newman. 2017. The evolutionary origin of digit patterning. *EvoDevo* 8(1):21.

Stewart, T. A., J. B. Lemberg, N. K. Taft, et al. 2020. Fin ray patterns at the fin-to-limb transition. *Proceedings of the National Academy of Sciences USA* 117(3):1612–1620.

Stoltzfus, A. 2021. *Mutation, Randomness, and Evolution*. New York: Oxford University Press.

Tanaka, Y., H. Kudoh, G. Abe, S. Yonei-Tamura, and K. Tamura. 2021. Evo-devo of the fin-to-limb transition. In *Evolutionary Developmental Biology: A Reference Guide*, edited by L. Nuño de la Rosa and G. B. Müller. Cham: Springer, 907–920.

Telford, M. J. and G. E. Budd. 2003. The place of phylogeny and cladistics in evo-devo research. *International Journal of Developmental Biology* 47:479–490.

ten Broek, C. M. A., A. J. Bakker, I. Varela-Lasheras, et al. 2012. Evo-devo of the human vertebral column: On homeotic transformations, pathologies and prenatal selection. *Evolutionary Biology* 39(4):456–471.

True, J. R. and S. B. Carroll. 2002. Gene co-option in physiological and morphological evolution. *Annual Review of Cell and Developmental Biology* 18:53–80.

Tschopp, P. and C. J. Tabin. 2017. Deep homology in the age of next-generation sequencing. *Philosophical Transactions of the Royal Society B: Biological Sciences* 372:20150475.

Tuomi, J. 1981. Structure and dynamics of Darwinian evolutionary theory. *Systematic Zoology* 30:22–31.

Tweedt, S. M. and D. H. Erwin. 2015. Origin of metazoan developmental toolkits and their expression in the fossil record. In *Evolutionary Transitions to Multicellular Life: Principles and Mechanisms*, edited by I. Ruiz-Trillo and A. M. Nedelcu. Dordrecht: Springer, 47–77.

van Amerongen, R. and R. Nusse. 2009. Towards an integrated view of Wnt signaling in development. *Development* 136(19):3205–3214,

Villegas, C., A. C. Love, L. Nuño de la Rosa, I. Brigandt, and G. P. Wagner. 2023. Conceptual roles of evolvability across evolutionary biology: Between diversity and unification. In *Evolvability: A Unifying Concept in Evolutionary Biology?* edited by T. F. Hansen, D. Houle, M. Pavlicev, and C. Pélabon. Cambridge, MA: MIT Press, 35–54.

Von Dassow, G. and G. M. Odell. 2002. Design and constraints of the *Drosophila* segment polarity module: Robust spatial patterning emerges from intertwined cell state switches. *Journal of Experimental Zoology (Mol Dev Evol)* 294B(3):179–215.

Wagner, A. 2011. *The Origins of Evolutionary Innovations: A Theory of Transformative Change in Living Systems*. New York: Oxford University Press.

Wagner, G. P. 2000. What is the promise of developmental evolution? Part I: Why is developmental biology necessary to explain evolutionary innovations? *Journal of Experimental Zoology (Mol Dev Evol)* 288:95–98.

Wagner, G. P. 2001. What is the promise of developmental evolution? Part II: A causal explanation of evolutionary innovations may be impossible. *Journal of Experimental Zoology (Mol Dev Evol)* 291:305–309.

Wagner, G. P. 2007. The developmental genetics of homology. *Nature Reviews Genetics* 8(6):473–479.

Wagner, G. P. 2014. *Homology, Genes, and Evolutionary Innovation*. Princeton: Princeton University Press.

Wagner, G. P., C.-H. Chiu, and M. Laubichler. 2000. Developmental evolution as a mechanistic science: The inference from developmental mechanisms to evolutionary processes. *American Zoologist* 40:819–831.

Wagner, G. P., E. M. Erkenbrack, and A. C. Love. 2019. Stress-induced evolutionary innovation: A mechanism for the origin of cell types. *BioEssays* 41(4):1800188.

Wagner, G. P., K. Kin, L. Muglia, and M. Pavlicev. 2014. Evolution of mammalian pregnancy and the origin of the decidual stromal cell. *International Journal of Developmental Biology* 58:117–126.

Wagner, G. P. and V. J. Lynch. 2010. Evolutionary novelties. *Current Biology* 20(2):R48–R52.

Wagner, G. P. and J. Zhang. 2011. The pleiotropic structure of the genotype-phenotype map: The evolvability of complex organisms. *Nature Reviews Genetics* 12:204–213.

Wake, D. B. and A. Larson. 1987. Multidimensional analyses of an evolving lineage. *Science* 238:42–48.

Wallace, B. 1986. Can embryologists contribute to an understanding of evolutionary mechanisms? In *Integrating Scientific Disciplines*, edited by W. Bechtel. Dordrecht: M. Nijhoff, 149–163.

Wasserman, G. D. 1981. On the nature of the theory of evolution. *Philosophy of Science* 48:416–437.

Weisberg, M. 2013. *Simulation and Similarity: Using Models to Understand the World*. New York: Oxford University Press.

Weisblat, D. A. and D.-H. Kuo. 2014. Developmental biology of the leech *Helobdella*. *International Journal of Developmental Biology* 58:429–443.

West-Eberhard, M. J. 2003. *Developmental Plasticity and Evolution*. New York: Oxford University Press.

Wilkins, A. S. 2002. *The Evolution of Developmental Pathways*. Sunderland, MA: Sinauer Associates.

Wimsatt, W. C. 1986. Developmental constraints, generative entrenchment and the innate-acquired distinction. In *Integrating Scientific Disciplines*, edited by W. Bechtel. Dordrecht: Martinus Nijhoff, 185–208.

Wimsatt, W. C. 1997. Aggregativity: Reductive heuristics for finding emergence. *Philosophy of Science* 64:S372–S384.

Wimsatt, W. C. 2007. *Re-engineering Philosophy for Limited Beings: Piecewise Approximations to Reality*. Cambridge, MA: Harvard University Press.

Winther, R. G. 2021. The structure of scientific theories. In *The Stanford Encyclopedia of Philosophy*, edited by E. N. Zalta. https://plato.stanford.edu/archives/spr2021/entries/structure-scientific-theories/.

Yang, A. S. 2001. Modularity, evolvability, and adaptive radiations: A comparison of the hemi- and holometabolous insects. *Evolution & Development* 3(2):59–72.

Acknowledgments

This book represents a condensation, elaboration, and synthesis of two decades of work on different conceptual issues arising at the junction of evolution and development. I am grateful to numerous colleagues who have patiently helped me formulate and analyze these issues along the way. To list only those I remember would reveal the poverty of my memory in documenting all their assistance. You know who you are. Two anonymous referees provided insightful comments and suggestions that improved the book. I also appreciate the input of my lab group over the past few years (Amanda Corris, Max Dresow, Lauren Wilson, and Yoshinari Yoshida) who gave last-minute comments and caught a variety of infelicities and typos. In particular, I thank my collaborators who have worked with me on material scattered throughout the book: Ingo Brigandt, James DiFrisco, Laura Nuño de la Rosa, Michael Travisano, Daniel Urban, Cristina Villegas, Günter Wagner, and Yoshinari Yoshida. My apologies to those I could not cite due to space constraints. I happily acknowledge the publishers of my papers from which some of the material contained herein was adapted: Elsevier, MIT Press, Oxford University Press, Routledge, Springer, University of Chicago, and Wiley. As Director of the Minnesota Center for Philosophy of Science, I have been ably assisted by Janet McKernan for many years. Without her, many (most) things would simply not happen. Two grant projects provided partial support for the work on this book, both funded by the John Templeton Foundation: "Integrating genetic and generic explanations of biological phenomena" (46919) and "The creative role of stress in development and evolution" (61329). (The viewpoints expressed in this book are the author's and not those of the John Templeton Foundation.) Support from the University of Minnesota came in the form of several Imagine Fund Annual Awards, the CLA Scholar of the College Award, the John M. Dolan Professorship, and the Winton Chair in the Liberal Arts. The most essential support came from my family. Lolene has been an unparalleled partner and the "better half" of the deal for more than twenty-five years. Our four larval forms – Nathanael, Caleb, Alethea, and Gideon – have largely metamorphosed, though dispersal is still in progress. You have been supportive beyond reason as I have worked on these topics for two decades. I couldn't imagine a better home team and wouldn't have come half as far without you: *Gratias vobis ago.*

For Rudy: a mentor first and forever a friend

Rudolf A. Raff (1941–2019)

Philosophy of Biology

Grant Ramsey

KU Leuven

Grant Ramsey is a BOFZAP research professor at the Institute of Philosophy, KU Leuven. His work centers on philosophical problems at the foundation of evolutionary biology. He has been awarded the Popper Prize twice for his work in this area. He also publishes in the philosophy of animal behavior, human nature, and the moral emotions. He runs the Ramsey Lab (theramseylab.org), a highly collaborative research group focused on issues in the philosophy of the life sciences.

Michael Ruse

Florida State University

Michael Ruse is the Lucyle T. Werkmeister Professor of Philosophy and the Director of the Program in the History and Philosophy of Science at Florida State University. He is Professor Emeritus at the University of Guelph, in Ontario, Canada. He is a former Guggenheim fellow and Gifford lecturer. He is the author or editor of over sixty books, most recently *Darwinism as Religion: What Literature Tells Us about Evolution; On Purpose; The Problem of War: Darwinism, Christianity, and their Battle to Understand Human Conflict;* and *A Meaning to Life.*

About the Series

This Cambridge Elements series provides concise and structured introductions to all of the central topics in the philosophy of biology. Contributors to the series are cutting-edge researchers who offer balanced, comprehensive coverage of multiple perspectives, while also developing new ideas and arguments from a unique viewpoint.

Cambridge Elements ☰

Philosophy of Biology

Elements in the Series

The Metaphysics of Biology
John Dupré

Facts, Conventions, and the Levels of Selection
Pierrick Bourrat

The Causal Structure of Natural Selection
Charles H. Pence

Philosophy of Developmental Biology
Marcel Weber

Evolution, Morality and the Fabric of Society
R. Paul Thompson

Structure and Function
Rose Novick

Hylomorphism
William M. R. Simpson

Biological Individuality
Alison K. McConwell

Human Nature
Grant Ramsey

Ecological Complexity
Alkistis Elliott-Graves

Units of Selection
Javier Suárez and Elisabeth A. Lloyd

Evolution and Development: Conceptual Issues
Alan C. Love

A full series listing is available at www.cambridge.org/EPBY.